THE SINO-AMERICAN JOINT COMMISSION ON RURAL RECONSTRUCTION

Twenty Years of Cooperation
for Agricultural Development

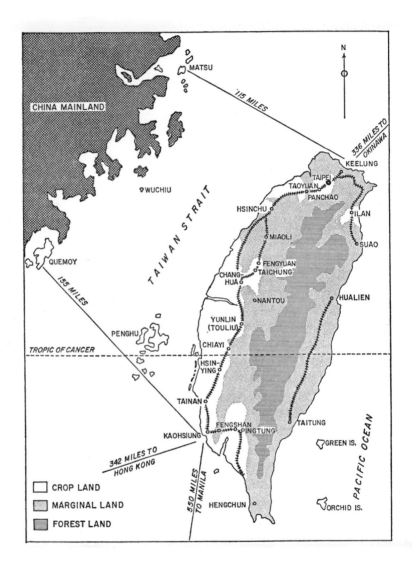

THE SINO-AMERICAN JOINT COMMISSION ON RURAL RECONSTRUCTION

Twenty Years of Cooperation for Agricultural Development

BY T. H. SHEN

CHAIRMAN, THE SINO-AMERICAN
JOINT COMMISSION ON
RURAL RECONSTRUCTION

CORNELL UNIVERSITY PRESS

ITHACA AND LONDON

International Standard Book Number 0-8014-0572-6
Library of Congress Catalog Card Number 72-118181

PRINTED IN THE UNITED STATES OF AMERICA
BY COLONIAL PRESS INC.

Foreword

To all who are concerned with the pressing and complex problems of agricultural and economic development throughout the world, I commend this book by Dr. Shen.

The results of the work of the Joint Commission on Rural Reconstruction in Taiwan over the past twenty years speak for themselves. They are in a word phenomenal.

Taiwan is the only developing nation that has reached the "take-off" stage of economic growth and no longer needs or asks to be included in the formal USAID program. It is now self-sufficient in food, has large exports of several commodities, and is now sending agricultural specialists to other countries to help them.

With quiet pride, Dr. Shen tells of the many successes of the JCRR program in Taiwan, but he also describes some of the failures and the reasons for them. The former far outnumber the latter, but both will be interesting and useful to other developing countries. If they studied the structure and the operation of JCRR, they might be able to adapt some of the ideas to speed their own agricultural and economic development.

Dr. Shen is the only man who could have written this book from personal experience because he is the only one left who has been with JCRR since its beginning. The book constitutes an important, clear-sighted history of a unique governmental invention and it is one more testimony of the fine leadership Dr. Shen and his associates are exerting in agricultural and rural development. It testifies also to the long and warm relations between him and Cornell University that reach back to the time he was a student and to the Cornell–University of Nanking project in the 1920's.

In addition to learning about the activities of JCRR, the reader will get Dr. Shen's underlying message loud and clear that expanding populations in the years ahead will make even greater demands for food and fiber and that a strong world agriculture will be a potent factor in world peace.

CHARLES E. PALM
Dean of the New York State
College of Agriculture at
Cornell University

Ithaca, New York
February 1970

Preface

This is a story about a unique organization, the Sino-American Joint Commission on Rural Reconstruction. It is unique in many respects: as a bilateral organization operating on a semi-autonomous basis; as an organization that is the first of its kind to prove the feasibility and effectiveness of binational technical cooperation; in the adaptability of its programs and projects covering all aspects of agriculture under widely different circumstances, at first on the Chinese mainland, and then on Taiwan.

My experience in Sino-American agricultural cooperation began with my participation in a crop improvement program sponsored jointly by the Plant Breeding Department of Cornell University and the University of Nanking in 1926 when I was a student at Cornell. I continued to take part in this program after I returned to China to join the faculty of the University of Nanking. For over sixteen years, from 1934 to 1950, I was connected with the Chinese National Agricultural Research Bureau, first as chief technician, then as Deputy Director, and finally as Director.

When the Chinese and American governments formed in 1946 a China–United States Agricultural Mission to study Chinese agricultural conditions and to make recommendations for the development of Chinese agriculture in the postwar years, I was a member of the Mission appointed by the Chinese government. From 1948 to the present time I have been a Chinese commissioner of JCRR, serving concurrently as JCRR chairman since 1964.

The first expression of my interest in Sino–American agricultural cooperation was an article I wrote entitled "How Can America Best Help China?" which was published by *Millard's Review* in July, 1920. In that essay I voiced the hope that the United States as the most advanced republic would help the young republic across the Pacific to develop her vast potential, especially her mineral and agricultural resources. Whether the realization of this dream in the 1920's instead of a quarter of a century later would have altered the history of China is now an academic question. It is a question, however, that the reader of this book may often ask himself.

In my preparation of the manuscript, a number of my JCRR colleagues have generously rendered me assistance. Those who helped me with the first draft are Mr. Teng Hsien-jen on water control and utilization; Dr. H. T. Chang, Dr. Chiu Ren-jong, Dr. Gideon T. W. Lew, Mr. Hsi Lien-chi, Dr. Puh Yen-sun, Mr. Wong Chun-muh, Mr. Horng Wei-huai, Mr. Chao Chih-kang, and Mr. Peng Tien-song on plant industry; Mr. Chen Tung-pai on fisheries; Dr. Robert Lee on animal industry; Mr. Yang Chi-wei on forestry; Messrs. Liu Ching-po and Sheng Tse-cheng on soil conservation; Dr. Hsu Shih-chu on rural health and family planning; Messrs. Yang Yu-kun and Chen Chin-wen on agricultural extension; Dr. Chen Hsing-yiu, the late Mr. Chuang Wei-fan, and Mr. Ho Wei-ming on agricultural marketing and export; Mr. Hsu Chien-yu on agricultural credit; Dr. Wang You-tsao on economic studies; Mr. King Yang-kao on the Chinese Agricultural Technical Mission to Vietnam; and Mr. Chang Hsuin-shwen on Chinese agricultural teams to the African countries. Commissioners Gerald H. Huffman and Y. S. Tsiang read Part I of the manuscript and both offered valuable suggestions for its improvement.

Professor Durham S. F. Chen, JCRR consultant, went over the entire manuscript in its draft form and made helpful suggestions. So did Professor William B. Ward of Cornell University, an expert in agricultural information whose views command both respect and assent. Miss Chi Tung, my secretary, has typed the manuscript. To all my colleagues and friends, who have shown infinite patience in helping me to write this book, I owe a deep debt of gratitude. But I alone am responsible for all statements of facts and expressions of views and judgments.

T. H. SHEN

Taipei, Taiwan
Republic of China
April 1970

Contents

Illustrations

MAP

PHOTOGRAPHS

following page 110

CHARTS

Tables

Abbreviations

AID	Agency for International Development
CATM	Chinese Agricultural Technical Mission
CIECD	Council for International Economic Cooperation and Development
ECAFE	Economic Commission for Asia and the Far East
ESB	Economic Stabilization Board
FA	Farmers' association
FAO	Food and Agriculture Organization, United Nations
MSA	Mutual Security Agency
NT$	New Taiwan dollars
NTU	National Taiwan University
PDAF	Provincial Department of Agriculture and Forestry
SAFED	Sino-American Fund for Economic and Social Development
SAU	Small agricultural unit

THE SINO-AMERICAN JOINT COMMISSION ON RURAL RECONSTRUCTION

Twenty Years of Cooperation for Agricultural Development

Introduction

Initially created as a nonpermanent Sino-American agency for the postwar rural reconstruction of China, the Joint Commission on Rural Reconstruction (JCRR) has helped more than 95 per cent of the rural people in Taiwan to live a better life. They enjoy a higher standard of living than their compatriots on the mainland under Communist rule. It has also helped the country make significant advances in agriculture, on the basis of which industry can develop and progress. That is how in less than twenty years the economy of the island, at first predominantly agricultural, has now become a mixed agricultural and industrial one, with industry assuming greater importance as time goes on.

With the transfer of JCRR to Taiwan and the reactivation of the United States economic aid to the Republic of China in this island province after the fall of the mainland to the Communists in 1949, JCRR has served as the agricultural arm of the United States Mission to China. This mission has been known by a succession of names, ECA, FOA, MSA, ICA, and, now, AID. Whereas the Mission devoted its attention mainly to industrial development projects, JCRR has worked on rural and agricultural development. This division of labor between the two agencies, one purely American and the other a bilateral enterprise, is a practical demonstration of how the technicians of two friendly countries can work side by side in a harmonious and fruitful relationship for the attainment of a common objective—the economic development of the less developed partner.

JCRR programs and projects were carried out first on the Chinese mainland, then in Taiwan, and, in later years, on the offshore island of Quemoy. From October, 1948, to November, 1949, when JCRR was still on the mainland, was a time of great uncertainty in view of the rapidly deteriorating political and military situation and of the run-away inflation. Under such trying circumstances, the JCRR commissioners and technicians continued to do their best to implement one project after another. Some of the projects such as the Tungting Lake Dike Repair were successfully completed before the fall of the mainland, but others had to be abandoned only half finished.

The peaceful and orderly environment in which JCRR has been able to operate after coming to Taiwan has been an important factor in the success of its programs. However, the results achieved on Quemoy seem to prove that, given the necessary determination and stamina, projects for rural improvement are possible even under the most adverse conditions. This island of the Republic of China has been under constant threat of invasion by the Chinese Communist forces poised on the mainland coast a few miles away and has been subjected to frequent artillery bombardments for a decade.

The projects are concerned with either technological improvements or institutional changes and readjustments. The former includes the introduction of new and improved varieties, better cultural methods, the application of more fertilizers in the right amounts and at the right times, the use of pesticides and irrigation water, and other technological innovations. The latter has to do with reform programs, of which the most noteworthy have been the implementation of the Land-to-the-Tiller Act, the reorganization of the farmers' associations, the irrigation associations, and the fishermen's associations, and the improvement of marketing procedures and facilities. By means of this two-pronged approach, agricultural production has been boosted and agricultural development stimulated in Taiwan.

Crops have been the chief concern of JCRR programs as well as of agricultural planning for Taiwan, with livestock coming second, fisheries third, and forestry last. The relative importance of each of these categories in the total agricultural production, which grew from US$676.9 million in 1952 to US$1,069.7 million in 1967, has changed only slightly from year to year. However, crops have always taken precedence. They represented 69 per cent of the total agricultural production in Taiwan in 1952, 64 per cent in 1961–1964, and 63 per cent in 1967. The slight drop in the relative importance of crops in total agricultural production has been compensated for by an increase in livestock production. It represented 16 per cent of total production in 1952, gradually rose to 21 per cent in 1957–1960, and finally to 22 per cent in 1967. Fisheries remained comparatively stable in relative importance, constituting about 9 per cent of the total agricultural production throughout the years. The relative position of forestry production did not change much either, accounting for some 5 or 6 per cent of the total in 1952–1967.

Throughout the 1950's the Joint Commission devoted most of its attention to the development of the western part of Taiwan. Here the fertile lands and the social and economic conditions favor the implementation of projects. Furthermore, the population is much denser than on any other part of the island. After accomplishing much for the farming people on the vast plains of the western coast, JCRR extended its operations to the poorer regions in eastern Taiwan and to the aborigines living in mountain areas. At the same time, attention was directed to the development of the offshore island of Quemoy. Originally a barren spot where scarcely anything could be produced, it has been turned into a thriving community capable of producing a number of farm products. Finally, JCRR in the past few years has also diverted a part of its attention and technical manpower to assisting friendly nations in Southeast Asia and Africa.

Since JCRR has only a small group of well-trained and experienced technical staff, its projects must be carried out in cooperation with other agricultural agencies, both public and private.

In the twenty years of its existence, JCRR has experienced both frustrations and successes. It has had to face serious difficulties and labor under many disadvantages, among which have been the rampant inflation, social chaos, economic dislocation, and worst of all, a fast deteriorating political and military situation during its mainland days in 1948 and 1949. The inevitable visitations of nature, such as typhoons and floods in Taiwan, have taken their toll. Moreover, JCRR, a non-permanent setup from the very beginning, might have been terminated at any time if it had not measured up to the expectations of the Chinese and American governments.

Accomplishments may be attributed to a variety of factors. The most important ones are a stable political and social order in which projects can operate, a progress-oriented government, a literate industrious and thrifty farming population, a steady flow of United States economic aid up to 1965, the flexibility of programs and projects, complete freedom from political pressure or other forms of outside influence, the happy choice of Chinese and American commissioners, and a real concern for the urgent needs of the farmers, whose views are solicited and are taken into consideration in the formulation of policy. Last but not least is the spirit of cooperation that has been consistently manifested by the Chinese and American staff members.

As the only one of the original five commissioners, three Chinese and two American, the writer has for the past twenty years held the position of commissioner and, since 1964, that of chairman. He has had the unique opportunity of observing the operations of JCRR from the beginning up to the present and of helping to formulate its policies and shape its programs. It is from this vantage point that he proposes to examine the JCRR record from its formative days on the mainland to the phasing out of United States economic aid in 1965 and to date.

The book is divided into four parts. Part I traces the origins of JCRR, outlines its policies, describes its organizational pattern and makes a brief analysis of its budget. Part II reviews agricultural rehabilitation and reforms carried out in Taiwan in the early 1950's, and the plans for new development. Part III details the major projects that have been implemented in Taiwan to demonstrate the JCRR approach to problem-solving. Included here also are chapters on agricultural development projects on Quemoy, in Vietnam, and in African countries which the Republic of China assists under its Vanguard Program. Part IV evaluates the JCRR effort throughout the last twenty years. It examines some of the unfinished tasks and points out wherein JCRR has failed.

This book, especially Parts II and III, is complementary to two previous works, *Agricultural Resources of China* and *Agricultural Development on Taiwan since World War II,* both published by the Cornell University Press. The reader is referred to the other works for background information and earlier statistical material. In the present study the material has been brought up to date.

In summary, it has been an unalterable principle of JCRR policy to help the great masses of small farmers raise their agricultural productivity and better the living conditions for them and their families. This work will show just how JCRR has shaped its programs and projects to meet the real needs of the farmers under the swiftly changing circumstances. It is hoped that this story will not only be of academic interest, but will also assist those who are actively engaged in finding ways and means to boost agricultural production and better the farmers' living conditions in their own countries through international technical cooperation.

PART I. BACKGROUND AND ORGANIZATION

1. Origins of the Joint Commission

Some developing countries have been exploring the possibility of creating bilateral agencies modeled after the Sino-American Joint Commission on Rural Reconstruction to administer the agricultural development programs. Though views on this question have been exchanged for quite some time, no concrete action has yet resulted. One obstacle has been the fear of the Asian nations that any such bilateral organization might be dominated by America as the senior partner. Most countries at the receiving end of United States aid tend to be suspicious of United States motives, especially if the plan for mutual cooperation originates in Washington.

However, no such fear ever occurred to the Chinese during the discussions that led to the formation of JCRR in the immediate postwar years. Indeed, it was the Chinese themselves who had made the proposal for a Sino-American joint agency to rehabilitate China's agriculture and promote its further development after eight years of war with Japan. After a century's experience of American friendship, the Chinese had not the slightest fear of American domination in any joint enterprise. They recalled with pride how Chinese and Americans had cooperated as equal partners to administer the Sino-U.S. China Foundation for the development of China's education. In the agricultural field, they had the example of the crop improvement program under the joint sponsorship of Cornell University and the University of Nanking from 1925 to 1931. This program, in which the writer was privileged to participate as a professor in the University of Nanking, foreshadowed in a way future agricultural technical cooperation between China and the United States. It laid a good foundation for crop breeding in China.

China–United States Agricultural Mission

In October, 1945, the Chinese government formally presented to the American government a proposal for technical collaboration in agriculture between the two countries. This eventually led to the creation of a joint China-United States Agricultural Mission whose task was to outline a broad and comprehensive program for agricultural development in China and to suggest the type and form of public service necessary for its implementation. The Chinese Ministry of Agriculture and Forestry appointed the writer in May, 1946, as its representative to draw up with Owen L. Dawson, his American counterpart, a program and itinerary for the Mission and to work out a set of recommendations on behalf of the Chinese to be submitted to the Mission for consideration.

In an exchange of letters between President Harry S Truman and President Chiang Kai-shek, Truman stated his "firm belief that any plan for cooperation in economic development between our two countries should include agriculture, the major source of income for such a great proportion of China's population." He continued: "In the experience of the United States, agricultural improvement has been found so important in promoting security, producing industrial raw materials, providing markets for industrial products, and raising the level of living that we believe a successful national development cannot be assured unless the development of agriculture proceeds simultaneously with the development of other elements in the national economy."

After pointing out how "our own agriculture is already indebted to your country for valuable agricultural material which has been introduced into the United States" and how "we still have much to learn from Chinese agriculture," Truman emphasized that "a high level of living for the whole of China's population, which can hardly be achieved without a strong development of agriculture, is the necessary foundation for the achievement of results that will benefit both of our countries, including an expansion of complementary trade and the development of China's industrial program."

In reply, President Chiang Kai-shek expressed agreement that "any plan for economic development between our two countries should include agriculture." He said that he was "keenly conscious of the fact that unless and until Chinese agriculture is modernized, Chinese industry cannot develop; as long as industry remains undeveloped, the general economy of the country cannot greatly improve." He firmly believed that the Chinese and American "groups working together will succeed in evolving plans and projects which will prove beneficial to China as well as helpful to the development of economic and trade relations between our two countries."

In the meantime, the group of eleven Americans appointed by Truman had arrived in China. It was headed by Claude B. Hutchison, dean of the College of Agriculture and vice-president of the University of California, and included Raymond T. Moyer, head of the Far Eastern Division, Office of Foreign Agricultural Relations, U.S. Department of Agriculture, as deputy head and secretary. Their respective Chinese counterparts were Tsou Ping-wen, senior adviser to the Ministry of Agriculture and Forestry and the Ministry of Food, and the author, deputy director of the National Agricultural Research Bureau, Ministry of Agriculture and Forestry. Like the American members of the Mission, all the thirteen Chinese members were presidential appointees.

The Mission began work in late June, 1946. It held a series of conferences in Shanghai and Nanking with government officials, agricultural specialists, educators, business leaders, bankers, and others well informed on Chinese agriculture, its current problems and its relation to the national economy as a whole. Then the Mission divided into six groups on the basis of the problems and special commodities to be studied. During eleven weeks of travel in which they covered 8,000 miles, the group members studied at first hand (1) agricultural education, research, and extension facilities and procedures; (2) the production, processing, and marketing of agricultural products; and (3) the economic and technical problems associated with rural life and with the use of land and water resources for agricultural production.

In its report submitted to President Chiang Kai-shek and President Harry S Truman, the Mission stated that "it is fully convinced that agricultural production in China can be greatly increased by the application of modern scientific knowledge to the improvement of soils, crops, livestock, and farm equipment" and that "the income of farmers can be greatly enhanced and the present poverty of many rural communities reduced by improvements in land tenancy, farm credit, and agricultural marketing." It expressed the "hope that some of the recommendations can form the basis for future collaboration between the two countries." Of the ten recommendations made by the Mission, the most important are:

1. That increased emphasis be placed on the construction of chemical-fertilizer plants; on the development of irrigation; on the improvement of plants and animals and their protection from insects and diseases; on forestry to provide timber for construction and fuel; and on the production of fruits, vegetables, and livestock to improve diets and nutrition.

3. That means be found to provide adequate farm credit at low cost; to assist farmers in marketing their products; to improve the conditions of tenancy where serious tenancy problems exist; to advance as rapidly as possible land surveys, registration, and appraisal; and to enforce the provisions of the Land Law of 1946 with respect to the taxation of land.

5. That programs relating to general education, public health, sanitation, transportation, river conservancy, and flood control be advanced as rapidly as possible.

7. That nine strong centers of agricultural instruction, research, and extension be established at points from which they can serve all sections of the country, at which there should be a Regional College of Agriculture, a Regional Agricultural Experiment Station, a Regional Agricultural Extension Office, and a Regional Agricultural Library. The centers proposed are: Nanking, Peiping, Changchun, Lanchow, Wukung, Chengtu, Wuchang, Canton, and Taipei.

10. That the Ministry of Social Affairs consider action which might suitably be taken by the Government to guard against a rapid increase in the growth of population.

Of these five recommendations, all but the seventh have been put into practice in Taiwan by JCRR. It is interesting to note that all but the third and tenth have been implemented on the mainland by the Communist regime side by side with their own ruthless land reforms, in which landlords were liquidated by the thousands, and the traditional family farms were transformed into impractical communes where the people were forced to work like ants.

The report was published in May, 1947. It proved to be of value to the United States Congress in the next year in its deliberations on economic aid to China. One senator cited its recommendations during the debate on the China Aid Act on March 30, 1948. Furthermore, the valuable experience gained in dealing with certain aspects of rural reconstruction through the Chinese Mass Education Movement headed by Y. C. James Yen was also taken into consideration by the United States government. Yen suggested the setting up of a joint commission to administer a program of mass education and rural reconstruction, and specifically recommended that 10 per cent of the aid to China to be authorized in the China Aid Act, then under consideration by Congress, be earmarked for rural development.

Changing concepts of foreign aid were in part responsible for the creation of JCRR. A new concept, embodied in the Economic Cooperation Administration, had arisen to take the place of the old, which had been too much in the nature of relief. This newer idea of helping others to help themselves could well be applied to China for the rehabilitation of her rural areas. Moreover, as Communism was capitalizing on the poverty and miseries of the Chinese peasantry, it was felt that the most effective way to steal the Communist thunder would be to solve the agrarian problem through peaceful reforms carried out with the technical and financial assistance of China's wartime ally and traditional friend, the United States.

The China Aid Act (Public Law 472, 80th Congress) on April 3, 1948, authorized, in Section 407(a), the Secretary of State "to conclude an agreement with China establishing a Joint Commission on Rural Reconstruction in China, to be composed of two citizens of the United States appointed by the President of the United States and three citizens of China appointed by the President of China." The terms of reference given the Commission were to "formulate and carry out a program for reconstruction in rural areas of China, which shall include such research and training activities as may be necessary or appropriate for such reconstruction." To carry out the purposes of the Act, Congress authorized the appropriation of US$338,000,000, of which not more than 10 per cent was made available for rural reconstruction in China.

Following the conclusion of an Economic Aid Agreement between China and the United States on July 3, 1948, an exchange of notes on August 5 formally authorized the establishment of the Sino-American Joint Commission on Rural Reconstruction in China.

Appointment of Commissioners

With the stage thus set, Raymond T. Moyer and John E. Baker were appointed by the President of the United States and Chiang Monlin, Y. C. James Yen, and T. H. Shen by the President of China as commissioners.

Dr. Chiang Monlin, who was elected chairman of the Joint Commission at its first meeting held in Nanking on October 1, 1948, had been president of the National Peking University for many years and had also served as Minister of Education and Secretary General of the Executive *Yuan*. Dr. Yen was noted for his experiment in mass education at Ting Hsien in Hopei Province, a work in which he had spent thirty years. Dr. Moyer had been an agricultural missionary in Shansi Province in northern China for fifteen years and was later appointed chief of the Far Eastern Division, Office of Foreign Agricultural Relations in the U.S. Department of Agriculture. Dr. Baker had been a relief executive in China for many years and had served as adviser to the Chinese Ministry of Railways.

The writer had been a long-time friend of Dr. Yen and Dr. Moyer. When he was teaching at the University of Nanking, Dr. Yen invited him to Ting Hsien to see the mass education movement at first hand and to help plan its crop improvement program and reorganize its agricultural department. Dr. Moyer had been the author's classmate at Cornell University and had cooperated with him in the crop improvement program of the University of Nanking. While Dr. Moyer was head of the agricultural department of Oberlin School at Taiku, Shansi Province, the author had paid annual visits to him to see its crop experiments in the years before the war with Japan.

All three Chinese commissioners had studied at American universities and through personal contacts and firsthand observations in the United States, they had come to appreciate American ways of approaching problems. Both of the Americans had decades of experience in working with Chinese in all walks of life. Without the intimate and friendly cooperation which existed among the five commissioners, the joint structure might have caused divided counsel and discord.

Division of Labor among the Commissioners

There was division of labor among the commissioners from the very beginning. Aside from their joint responsibilities as the policy-making body of the bilateral organization, all commissioners are assigned certain specific duties. In the early days when there were five commissioners, Chairman Chiang Monlin was responsible for maintaining liaison with high-level national and provincial agencies and officials, general supervision over administrative and personnel matters relating to the Chinese staff, and co-supervision over the Commission's financial matters; Raymond T. Moyer, for maintaining liaison with United States government agencies and officials, general supervision over administrative and personnel matters relating to the American staff, co-supervision over the Commission's financial matters, and general supervision over the Fertilizer Division program; T. H. Shen, for general supervision over programs of the Agricultural Improvement, Land, Farmers' Organization, and Animal Industry Divisions; John E. Baker, for general supervision over programs of the Irrigation and Rural Health Divisions and general supervision over project audits and project reporting to ECA Washington; and Y. C. James Yen, for general supervision over the dissemination of public information and the utilization of public information facilities of JCRR.

In spite of changes in both the personnel and the number of commissioners as well as numerous changes in the organizational pattern of JCRR in the last twenty years, which are described in some detail in a later chapter, the principle of division of labor among the commissioners has been maintained intact.

2. Policies and Program

The terms of reference given to JCRR under the China Aid Act were broad, and the five Commissioners were faced with the challenging task of formulating specific policy lines.

Shortly before the formal inauguration of JCRR, it received from the Economic Cooperation Administration (ECA) in Washington a "Proposed Program" dated September 20, 1948, in which certain principles, programs, and procedures were suggested. This carefully thought-out document was not intended as a directive and, therefore, many of its suggestions were not acted upon. However, it proved to be a useful guide and provided the main tentative outline for the organization and work of the Joint Commission in the early period of its existence. This very first communication from ECA clearly showed that the United States was not trying to impose its will and that the Joint Commission was not expected to be a rubber stamp for Washington.

JCRR Policy

Having assumed office, the five Commissioners set out to formulate policies and an initial program and worked very hard at it. They held a series of meetings at which each one of them freely expressed his own views and with equal freedom commented on those expressed by his colleagues. One point to be noted is that the commissioners, who had different backgrounds, experience, and outlooks, stated only their individual views as experts and not as representatives of this or that country. Naturally, there were some sharp differences of opinion which, being honest and sincere, merely served to enhance mutual respect. The commissioners firmly believed that only through a candid exchange of views could they hammer out a practical program for the solution of China's rural problems.

A unique and continuing feature of the Commission meetings has been that no formal vote is ever taken on any question or proposal. Whenever agreement cannot be reached on an item of business, the matter is postponed, to be taken up again at subsequent meetings until there is unanimous agreement. If unanimity is not reached the question remains in abeyance. At first sight, this way of conducting business may seem to be contrary to the general practice of majority rule in a democratic assembly. But since the Joint Commission is composed of unequal numbers of Chinese and American commissioners, one group could always have a majority if votes were counted. The tradition of unanimity has assured adequate minority representation and, much more important, given the decisions of the Commission a weight and an impact that they would otherwise have lacked. A single dissenting voice can block the approval of a given project, unless those in favor of it can persuade the dissenter to change his mind. This odd method of transacting business in meetings of the Joint Commission, though seemingly cumbersome and time-consuming, has proved to be an important factor in creating a spirit of genuine cooperation in JCRR and contributing to its success.

A set of objectives and principles relating to programs and procedures was worked out at this time. The four basic ones which have guided JCRR operations throughout the last twenty years are:

1. There must be felt a need for JCRR services on the part of farmers themselves, who are after all the direct beneficiaries of the projects.

2. There must be a fair distribution of the benefits that are derived from the JCRR projects.

3. There must be a sponsoring agency qualified to undertake any given JCRR project and to utilize JCRR assistance effectively.

4. There must be a demonstration of feasibility of any particular JCRR project before undertaking its broad expansion.

It cannot be overemphasized that the requirement of a sponsoring agency to implement JCRR projects has been one of the best policy decisions made. The sponsoring agency is usually chosen from local organizations that have a good reputation for efficiency and honesty and enjoy the confidence of the local people. As the agency so chosen is part of the community in which the JCRR project is to be operated, it knows the local conditions better than an outsider.

Program Objectives in Taiwan

Taiwan's agricultural production, with the single exception of sugar, was largely for home consumption until about 1960. At this time several institutional changes facilitated industrial development and the export of agricultural products. Since then agricultural production has been aimed at not only providing food and fiber for domestic consumption, but also leaving a surplus for export, and supplying industry with raw materials. In this way, Taiwan's economy has gradually been transformed from a predominantly agricultural to a mixed agricultural and industrial one.

In the light of these developments, the Joint Commission re-examined its program objectives in 1963. After identifying the key problem areas, it set forth three basic production goals for subsequent programs:

1. To increase the production of food for domestic use at the average annual rate of 4.5 per cent in order to provide, at reasonably stable prices, the basic nutritional requirements for Taiwan's population.

2. To expand, diversify, and improve the quality of agricultural exports, including processed products, at the annual rate of 6.5 to 7 per cent over the next five years, as an added agricultural contribution to foreign exchange earnings and to the achievement of the overall national goal of economic self-support.

3. To provide an increasing supply of raw materials of agricultural origin for industrial processing and industrial expansion.

To achieve these basic JCRR program goals, the following supporting development goals were also agreed upon:

1. To further develop, conserve, and manage Taiwan's land and water resources, based upon a province-wide priority-determined development plan in order to increase agricultural output and reduce flood damage to the Island's communities and capital facilities; to develop other capital resources related to agricultural development as required.

2. To strengthen the operations of agricultural cooperatives and other marketing facilities, and improve the marketing and distribution system of both production requisites and commodities produced.

3. To further raise the level of organizational efficiency and coordination of those agricultural and rural institutions—research, education, extension, credit, rural health and producer associations—that have a direct demonstrable contribution to make to increased agricultural production and productivity.

With the phasing out of United States economic aid to China on June 30, 1965, another change in JCRR policy has been the gradual shift to projects of an innovative and developmental nature and the gradual transfer of more or less routine projects activities designed to support the government budget to appropriate governmental agencies. All through the last twenty years, the Joint Commission has endeavored to make its program dynamic, to meet the needs of the changing times and to boost agricultural production and better the livelihood of the rural people. Needless to say, agricultural development has indirectly contributed to industrial development as well.

3. Organization and Staff

The five commissioners, three Chinese and two American, constituted the governing body. It was they who made all the policy decisions and exercised general supervision over the implementation of JCRR supported projects with the assistance of a technical and administrative staff.

A number of changes have taken place in the composition of the Joint Commission in the last twenty years. Of the original commissioners, two have resigned, two died, and only the writer still remains at his job. Dr. Yen, who resigned in 1951, and Drs. Baker and Moyer, who completed their tours of duty in 1951 and 1952, were succeeded by Messrs. Chien Tien-ho, William H. Fippin, and Raymond H. Davis, respectively. Before their appointments as commissioners, all the three men had held various important positions. Chien was deputy director of the National Agricultural Research Bureau and then vice-minister of the Ministry of Agriculture and Forestry of the Chinese Government from 1940 to 1947 and head of the Agricultural Division of JCRR from 1948 to 1952. Fippin had been an agricultural officer of the ECA/Korea Mission before his transfer to Taiwan as head of the Farmers' Organization Division of JCRR in 1951. Davis had been a soil conservationist and former Agricultural Chief of the Military Government in Japan during the allied occupation.

After their tours of duty were ended, Fippin's in 1957 and Davis's in 1959, they were reassigned to posts in other countries. To take Davis's place, the American government appointed Mr. Clifford H. Willson, who had been director of the Technical Service of FAO, United Nations, and a member of the International Development Board for Iraq. However, since the United States Government never filled the vacancy left by Fippin, there were only four JCRR commissioners—one American and three Chinese—from Fippin's departure in June, 1957, to the death of Chiang Monlin in June, 1964.

After Dr. Chiang's demise, the writer was elected by the Joint Commission to succeed him as chairman. Simultaneously, another reduction in the number of commissioners was effected by an exchange of notes between the Chinese and American governments on August 31 and September 1, 1964; it was agreed that "it is the intention of the United States of America not to fill the vacancy existing in the number of American Commissioners of the Joint Commission on Rural Reconstruction provided for in the Agreement of August 5, 1948 on the understanding that it is the intention of the Government of the Republic of China not to fill the vacancy existing in the number of Chinese Commissioners provided for in the aforementioned Agreement." This exchange had the effect of limiting the number of American commissioners to one and that of the Chinese to two.

When Commissioner Chien retired in August, 1961, Y. S. Tsiang, who had been head of the miscellaneous crops department of the National Agricultural Research Bureau and later secretary-general of JCRR, stepped into his shoes. With Commissioner Willson's resignation for reasons of health, Gerald H. Huffman, who had been deputy director of the Federal Extension Service of the U.S. Department of Agriculture, was appointed by Washington to succeed him in June, 1962. He served until April, 1968, when he was reassigned to the AID Mission to Vietnam as Associate Director for Domestic Production.

At present, the three commissioners are Bruce H. Billings, who succeeded Huffman as representative of the United States government, and Y. S. Tsiang and the writer, representing the Chinese government.

Dr. Billings, the American Commissioner, who serves concurrently as science adviser to the American Embassy in Taipei, is a top physicist and has a broad knowledge of science and a deep interest in China. He was vice-president and general manager of laboratory operations of Aerospace Corporation, Los Angeles, California, before his appointment by President Lyndon B. Johnson as JCRR Commissioner.

Changes in Organizational Pattern

The changing times have necessitated readjustments and modifications in the programs and activities of JCRR. Organized on a subject-matter basis, the technical divisions have undergone more changes than the administrative offices.

The first important change took place soon after JCRR's removal to Taiwan in 1949. Three of the original four divisions set up in the previous year were abolished, and the Agricultural Production Increase Division, which was retained, was renamed the Agriculture Division. Four new divisions were created, those of land tenure reform, audio-visual education, irrigation, and rural health.

The second change, effected in February, 1950, resulted in the abolition of the Audio-Visual Education Division. The next change, effected in September, 1952, led to the replacement of the Executive Office by a Commission Secretariat headed by a Secretary-General. The number of technical divisions was increased to nine by breaking up the Agricultural Improvement Division into the Divisions of Plant Industry, Animal Industry, and Forestry, renaming the Land Division the Land Reform Division, the Fertilizer Distribution Division the Food and Fertilizer Division, and the Irrigation Division the Irrigation and Engineering Division, adding a Rural Economics Division, and retaining the Farmers' Organization and Rural Health Divisions.

In March, 1957, the number of technical divisions was cut from nine to the following seven: Plant Industry, Animal Industry, Forestry, Irrigation and Engineering, Rural Health, Rural and Land Economics (formerly Rural Economics), and a newly created Agricultural Extension Division. The Land Reform and Farmers' Organization Divisions were abolished because their objectives had been achieved.

In August, 1958, the number of technical divisions was again increased to nine. An Agricultural Credit Division was added, and the Farmers' Organization Division revived. The Rural and Land Economics Division was renamed the Rural Economics Division. In April, 1959, a Fisheries Division was created. In July, 1963, the Farmers' Organization and Agricultural Extension Divisions were amalgamated into a Farmers' Service Division, bringing the divisions once more to nine. Thus the JCRR adapts its structure to cope with the needs of the times.

As a result of various changes in the technical and administrative staffs of the Joint Commission, an organizational pattern has emerged (Chart 3-1).

Recruitment of Staff Members

Since JCRR is a bilateral organization, its staff members have consisted of American and Chinese experts, together with a number of supporting administrative personnel. The two nationalities have in effect complemented and supplemented one another. The policy has been not to employ American experts if an equally qualified Chinese expert is available. Moreover, the American experts help to train their Chinese counterparts and as soon as the latter can take over the job, the former return to the United States. In a few divisions such as plant industry, fisheries, and rural health, Americans have not been needed, except as short-term consultants.

Chart 3-1. Organization of JCRR, 1969

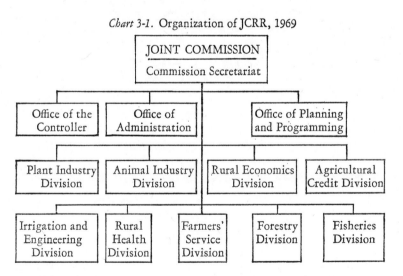

An overwhelming majority of the Chinese technical staff, who are all specialists in their own fields, have, like the Chinese commissioners, studied in American colleges and universities. Thus they are able to work hand in hand with their American colleagues, each of whom has been chosen on the strength of his technical competence for his assignment with JCRR.

The American members of the technical staff not only worked with their Chinese counterparts in the Joint Commission but also came into wide and direct contact with the staff members of sponsoring agencies of JCRR projects. In this way, their influence was felt in other organizations. But the limitation of the tenure of office of the American personnel to two years was a distinct disadvantage. Though an individual's tour of duty might be extended for a term, he was usually reassigned to another post at home or abroad at the end of either the first or the second term. Some of the Americans would have made still greater contributions to Taiwan if they could have worked two or four years longer.

In the first few years of JCRR's existence, the number of American personnel on its technical staff averaged eight to ten. It rose to thirteen from 1954 to 1963, not including short-term consultants on highly specialized topics. Since then the number of American technicians has been gradually reduced until at present only one American commissioner remains.

The Chinese staff members have been carefully selected on the basis of their training and previous experience and achievements. The heads of divisions are recommended by the commissioners and their appointment is approved with the unanimous consent of all the commissioners at meetings of the Joint Commission. The division heads recommend for the approval of the Joint Commission properly qualified individuals to serve as specialists in their respective divisions.

The initial selection of senior Chinese agricultural specialists by the Joint Commission presented a problem. Fortunately, the National Agricultural Research Bureau had recruited a group of well-trained and energetic young scientists. With the Bureau's consent, the Joint Commission invited ten of its senior specialists to join the JCRR staff; one of them, Y. S. Tsiang, later became a commissioner. From this nucleus, JCRR's agricultural staff grew as the evolving programs dictated. Since its removal to Taiwan, new staff members were recruited locally.

The JCRR staff members numbered only 40 when the Joint Commission started its operations with four technical divisions in 1948. But as the number of technical divisions gradually increased until it reached ten in 1959, there was a corresponding increase in personnel, reaching a maximum of 243 in 1962. The present nine technical divisions are manned by 105 technicians, with 90 supporting and administrative personnel.

Salary Scale

As the salaries of the American staff were paid by the Economic Cooperation Administration according to its own salary scale, they did not present a problem to the Joint Commission. But in the case of the Chinese staff, the matter was complicated by the excessively low salaries paid to government employees on the one hand and the runaway inflation in the immediate postwar years on the other. Before making any decision on this question, the Joint Commission took into consideration several factors, namely, that all JCRR salary payments would come from United States economic aid funds, that the Chinese staff members of JCRR did not form part of the Chinese civil service, and that JCRR was considered to be a temporary organization which could be closed at any time. In view of these facts, it was decided that the Joint Commission had authority to fix its own salary scale subject to the approval of the Premier. Then, a careful study was made of the prevailing salary scales of both government and private agencies. Finally, the following guiding principles were adopted: (1) the salaries paid to Chinese employees must provide a decent living; (2) a straight salary should be paid, without food and other allowances such as those received by government employees; (3) the salary should be for full-time work, and no JCRR employee should be allowed to draw pay for concurrent jobs in other institutions. By following these principles, the Joint Commission fixed a salary scale for its local employees 100 to 200 per cent higher than the government's at that time, which the Commission was convinced was too low.

The Joint Commission was also called upon to consider the pay for the Chinese commissioners. The chairman's salary finally agreed upon was equivalent to US$5,400 per annum and that of the two other

Chinese commissioners, US$4,200. In discussions over the question at Commission meetings, the American commissioners considered the figures to be much too low in view of the much higher pay they themselves were drawing (US$12,400 in January, 1950, gradually rising to US$24,500 in June, 1965, plus post allowance, and, depending on family status, separation allowance and education allowance, while the pay of the Chinese commissioners, including the chairman, has remained fairly constant over the years). But the Chinese commissioners took the view that a relatively lower pay for themselves would be good strategy for them to use to defend the comparatively higher salary scale they were setting for the Chinese staff. They also felt that as all JCRR funds came from the American tax-payers' pockets, a lower pay for themselves would ease their own conscience.

At this point it should be explained that the United States dollar was adopted as the standard for fixing the salary scale in the initial years because inflation was rampant at the time. As soon as the New Taiwan dollar had been stabilized, it became the medium of payment, beginning in August, 1951.

Although the salary scale fixed by the Joint Commission was sanctioned by the Premier, the late General Chen Cheng, some legislators strongly objected, in the first few years after the Joint Commission's transfer to Taiwan, to the relatively higher pay JCRR employees were receiving. To meet their objections Chairman Chiang Monlin took pains to explain to the legislators why it was necessary for JCRR to pay its employees somewhat better than those of other agencies. His arguments resulted in greater understanding among the critics, and the JCRR salary scale received the tacit approval of all parties concerned. With the matter thus settled, the Joint Commission invited all members on the Economic Committee of the National Legislature to make a personal inspection of some projects then in progress. Since then there have been no complaints about the salary scale.

As a result of the general economic progress made in Taiwan in recent years, many private industries have been paying their employees better than JCRR. At the same time, the need for a higher salary scale for government employees has become evident. In a speech before the National Legislature prior to the 1967 salary increase for government employees, Vice-President-Premier C. K. Yen suggested that the government pay scale should be adjusted more or less according to the principles followed by JCRR.

Since the salary scale of JCRR is much lower than that of many foreign countries, the Joint Commission has lost the services of specialists to foreign countries; it lost twenty-three in the 1959–1967 period. The organizations that have recruited JCRR personnel through the offer of positions, at salaries five to eight times higher than those of JCRR are the World Bank, the Asian Development Bank, the International Rice Research Institute, FAO, the United Nations Economic Commission for Asia and the Far East (ECAFE), and other international institutions. Fortunately, equally capable specialists have come up from the ranks to replace those who have left. The implementation of JCRR programs over the years has created a pool from which new blood can be drawn. There is no denying, however, that Taiwan is feeling the pinch of brain drain as many other countries do.

During their visit to Taiwan in September, 1967, Donald F. Hornig, special assistant on science and technology to President Lyndon B. Johnson, and his team of experts discussed the brain drain with JCRR officials. The writer pointed out that it constituted a serious problem in agriculture, with the government organizations suffering even more than JCRR because of their lower pay scale. Robert Lee, chief of the JCRR Animal Industry Division, emphasized, however, that though JCRR had not suffered so much brain drain as other agricultural organizations in Taiwan, this was not due to the higher pay JCRR personnel had been drawing, but rather to the fact that the JCRR technical staff had been able to work in a favorable environment, free from bureaucratic control and red tape, and free to innovate and experiment with new ideas and methods.

JCRR Relationship with Other Agencies

Though JCRR is an autonomous and semi-independent organization, it is subject to a certain degree of control by the Chinese and American governments. With authority delegated by the director of his parent organization in Washington, now known as the Agency for International Development (AID), the director of the AID Mission to China has exercised policy direction and fiscal control over JCRR. The Joint Commission has acted as the agricultural arm of the AID China Mission, including in its functions those normally carried out by the Food and Agricultural Division of AID missions to other countries.

On the Chinese side, JCRR is subordinate to the Cabinet and subject to the direction and supervision of the Premier. Aside from this, the relationship between JCRR and other Chinese government agencies is not clearly set forth in any document. Being unique and bearing no resemblance to any other organization, JCRR's actual relationship with other Chinese institutions is rather indeterminate and depends to a large extent on human factors. The JCRR Chairmanship is generally considered to be of ministerial rank.

The relationship between JCRR and the Ministry of Agriculture and Forestry in the mainland days (1948–1949) was not officially defined. There might have been areas in which their activities and authority duplicated one another or even came into conflict. The writer, being a paid director of the National Agricultural Research Bureau, a subordinate organ of that Ministry, and concurrently a JCRR commissioner without pay, was placed in a most delicate and difficult position. Fortunately, the Minister lent his authority to encourage the Bureau to work in close cooperation with JCRR. With financial aid from JCRR, the Bureau carried out cooperative projects by making the best use of its own specialists and, in this way, made important contributions to agricultural production.

Following the removal of the seat of government to Taipei in 1949, there was a general reduction in the number of personnel of the central government as well as a reshuffling of the Ministries. The Ministry of Agriculture and Forestry was abolished; its functions were transferred to the Ministry of Economic Affairs, under which a Department of Agriculture was created to take charge of matters relating to agricultural development. At the same time, the National Agricultural Research Bureau, of which the writer was director, came under the jurisdiction of the Ministry of Economic Affairs. In 1950, the writer resigned from the Bureau to devote his full time to JCRR, but continued to maintain cordial relations with his former colleagues and superiors in the Ministry.

The year 1953 saw the beginning of the First Four-Year Economic

Development Plan and the establishment of the Economic Stabilization Board which had the Premier as chairman. With the chairman
of JCRR and the writer, in his capacity as a JCRR commissioner, serving as two of the Board members, the Joint Commission developed

Chart 3-2. Relationship of JCRR with government agencies

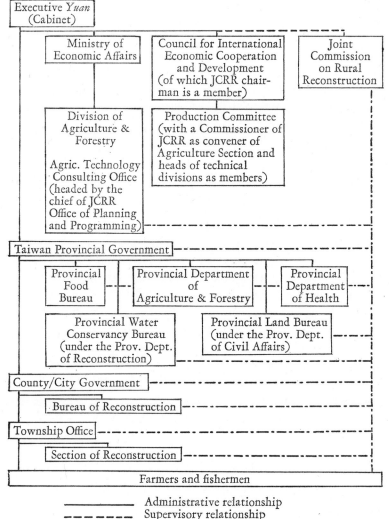

more functional relations with the government organizations in planning and implementing the agricultural programs of the successive Four-Year Plans. Those relations will be discussed in greater detail in the section "Joint Program Planning" in Chapter 4.

The relationship of JCRR with the agricultural agencies of the provincial government has been friendly since the mainland days, because all parties work for the same objective. JCRR renders technical and financial assistance to projects of the provincial government agencies. For example, the Department of Agriculture and Forestry of the Taiwan Provincial Government, which carries the burden of the agricultural field work in Taiwan, is one of the major recipients of JCRR aid funds. Thus the relationship between the Department and JCRR has always been most cordial. In fact two of JCRR's key staff members served as commissioners of the Department—Y. K. King from 1954 to 1962, and H. T. Chang from late 1962 to 1965. Similarly, Teng Hsien-jen of JCRR served as director of the Water Conservancy Bureau of the Taiwan Provincial Government.

JCRR's Life Expectancy

The original exchange of notes between the United States and China on August 5, 1948, provided for the establishment of JCRR and its continuation only until June 30, 1949. Although a subsequent exchange of notes on June 27, 1949, extended the life of JCRR for another twelve months, there was uncertainty in the early years of JCRR's existence as to how long it would continue to function. The Americans considered it a temporary organization, but the Chinese commissioners held a different view. They knew only too well that rural reconstruction was an immense task which could not be solved in one or two years.

As early as 1949, four private associations in Taiwan addressed a joint letter to the United States Department of State expressing the hope that JCRR would be made into a permanent institution. A Report of the Committee on Foreign Affairs of the U.S. House of Representatives, prepared at the time of the passage of the 1954 Mutual Security Act, contained these observations:

In other countries the emphasis is on, and usually limited to, the provision of American technical and financial assistance without direct responsibility for its effective use. In Formosa the principle of "jointness" is practiced at all levels—formulating policies, working out programs and procedures, and assuming responsibility for the results. The constant intent is to give the Chinese themselves the role of an active participant rather than a passive recipient of American assistance. . . .

The JCRR has been operated on the principle that well organized farmers' organizations are an essential factor toward accomplishing any rural reconstruction program. The objective of "a federated system of multi-service farmers' cooperatives that is organically strong, democratically controlled, efficiently operated and financially secure, alert and responsive to the interests and needs of members, and constantly seeking to increase its effectiveness and improve its services" has been sought through farmers' associations and rural youth organizations comparable to our 4-H Clubs. Recognizing that quality of staff is of highest importance, the JCRR has supported a number of projects designed to train directors, general managers and staffs of farmers' associations. The members talked to some of these agricultural leaders and were impressed with their ability and understanding approach.

The JCRR's contributions should not be measured alone in cold statistics of increased production. More significant, although less frequently recognized, is the training in leadership and democratic processes it has produced among the farmers and also among officials. Such values are bound to be reflected in the political evolution of the country.

The advantages derived from this approach to technical cooperation have been amply demonstrated. It is for that reason that the committee recommends the continuation of the JCRR. More than that, it hopes that the officers responsible for carrying out the technical cooperation program will make every effort to establish and handle as much of its work as possible through a similar body in each of the countries where our programs are in operation.

An evaluation of JCRR's contributions to agricultural development and the decision to have these contributions continued even after the phasing out of United States economic aid to China in 1965 will be the subject of the last chapter.

4. Program Operation

To apply practical solutions to rural problems, JCRR often had to modify its policies to meet changes in the general situation and the requirements of the areas served. Starting with a program in crop improvement and irrigation, it has over the years gradually expanded its activities to include land reform, farmers' organization, animal husbandry, forestry, rural health, fisheries, agricultural credit, and rural economics. Benefiting by experience, the Joint Commission has developed a program procedure that is spelled out in great detail in Administrative Order No. 25, dated May 16, 1952.

JCRR projects are generally initiated by local agencies which request technical or financial assistance. Some projects of an innovative nature have been initiated by JCRR specialists, who enlist appropriate agencies as project sponsors.

The requests from sponsoring agencies for aid are reviewed by the Divisions concerned. In reviewing such requests, the Divisions take into consideration the following factors: (1) conformance with approved program objectives; (2) technical feasibility; (3) reliability of cost estimates; (4) availability of funds; and (5) availability of technical and administrative services required. The Divisions may then recommend either approval, modification, tabling, or rejection of the requests.

When a project recommended by a particular Division has some bearing on, or relation to, another Division or Divisions, it is passed on to them for comment before it is submitted to the Joint Commission for action. Each project recommended for approval must include a schedule of payment, which is approved along with the project.

However, this approval does not commit the Joint Commission to make payment unless the project progresses satisfactorily.

Of the 1,207 project applications received by JCRR between October 1, 1948 and February 15, 1950, the Joint Commission approved 304 and rejected the other 903. In the following years, from 1950 until the phasing out of United States aid to China on June 30, 1965, a total of 6,280 projects were approved by the Joint Commission. The sponsoring agencies came to understand the objectives and procedure of JCRR projects, and far fewer project applications were rejected in later years than during the initial period.

In granting subsidy for any project, JCRR requires the sponsoring agency to contribute matching funds commensurate with its financial ability, because it is felt that such contributions express determination on the part of the sponsor to make the project a success. In 1950–1965, the project beneficiaries contributed on the average 48 per cent of the cost of JCRR-supported projects.

The role played by JCRR in agricultural development has been on the whole that of a catalytic agent. By providing technical and some financial assistance, JCRR has been able to give direction to agricultural policies, encourage improvements in methods of implementation, motivate projects difficult to activate, and generate a spirit of self-help among the local agencies. In line with this principle, over 96 per cent of all JCRR-supported projects[1] have been sponsored by officially recognized agencies and organizations.

From the very beginning it was understood that JCRR could work with any level or unit of a sponsoring organization in carrying out JCRR projects. In its early years it preferred to make grants or loans to the lowest unit of an organization so that it could be held directly responsible for implementing the project and accounting for funds expended. In this way bureaucratic control was bypassed, red tape reduced to a minimum, and assistance extended to the end-users at the grass roots. This system has made it possible for government agencies to be of real service to the people. Though some Chinese and American observers have criticized the Joint Commission's preference for

[1] Y. S. Tsiang, *The Joint Commission on Rural Reconstruction in the Republic of China* (Taipei: JCRR, 1964), p. 5.

working with so many low-level agencies, the results have demonstrated the essential soundness of such an approach.

It has also been JCRR policy to remit project funds directly to the end-user agency by installments. The disbursement of each installment must be justified by a "certification of progress" submitted by the technical Division exercising surveillance over the project, and contributing technical assistance to it. JCRR has sometimes been criticized for being slow in disbursing funds, or over-cautious in making allotments. But, under the terms of a project agreement, every request for funds must be scrutinized with the utmost care, however big or small the project may be.

With the legal and moral obligation to insist on the effective and efficient use of all technical and financial assistance rendered, the Joint Commission holds that it is not how much assistance should be given, but for what.

Joint Program Planning

For the administration of economic planning in the last fifteen years, several agencies were created. First came the Economic Stabilization Board (ESB), with the Premier as chairman, in 1953; then the Agricultural Planning and Coordination Committee (APCC) under the Ministry of Economic Affairs, in 1958; and finally the Council for International Economic Cooperation and Development (CIECD), also with the Premier as chairman, in 1963. These agencies, which succeeded one another, have had charge since 1953 of the successive Four-Year Economic Development Plans.

A unit under the Economic Stabilization Board, known as Committee D, was responsible for the planning and coordination of agricultural programs. The writer, a JCRR commissioner, was appointed to serve concurrently as a member of ESB and convener of Committee D, which included in its membership the director of the Department of Agriculture of the Ministry of Economic Affairs, the commissioner of the Taiwan Provincial Department of Agriculture and Forestry, the director of the Taiwan Provincial Food Bureau, the director of the Taiwan Provincial Water Conservancy Bureau, JCRR commissioners, the president of the Taiwan Sugar Corporation, and the dean of the College of Agriculture of the National Taiwan University. It was Committee D's primary duty to coordinate the programs, operations, and budgets of all these agencies within the framework of their respective organizations and functions.

JCRR played an active role in planning and implementing the agricultural program of the economic development plans. A JCRR senior specialist served as its executive secretary, and four other specialists were on its secretarial staff. The various divisions also gave technical assistance and advice to the Committee.

When the Agricultural Planning and Coordination Committee replaced the ESB in 1958, the same officials who had been members of Committee D in the previous years were appointed by the Minister of Economic Affairs to be members of the newly created APCC.

Transformed from the Council for United States Aid when American economic assistance was about to be phased out, the present Council for International Economic Cooperation and Development is composed of cabinet ministers in charge of economic affairs, finance, communications, education, national defense, and home affairs and the JCRR chairman. CIECD has set up a Production Committee, under which there are an Agricultural Subcommittee and an Industrial Subcommittee. The Production Committee is headed by the Minister of Economic Affairs assisted by two deputies, one of whom serves as convener of the Agricultural Subcommittee and the other, convener of the Industrial Subcommittee. JCRR commissioner Y. S. Tsiang is both deputy head of the Production Committee and convener of the Agricultural Subcommittee. Besides providing secretarial personnel for the latter Subcommittee, JCRR staff members, both Chinese and American, continue to take part in its work.

In line with the Government's general policy on economic development in Taiwan, a set of objectives for the development of the agricultural sector has been formulated: (1) to bring about stabilization of the agricultural prices through the provision of an adequate supply of agricultural products to meet the needs of the civilian and military population; (2) to expand agricultural exports in order to increase foreign exchange earnings; (3) to reduce agricultural imports so as to effect a saving in foreign exchange; and (4) to improve the livelihood of the farmers and, simultaneously, to maximize the farmers' contributions to the national economy. This set of objectives has been shared by the planners of the agricultural Four-Year Plans and JCRR.

Before the phasing out of United States economic aid to China in 1965, JCRR prepared a program of its own within the framework of the current Four-Year Economic Development Plan for each fiscal year and, with this program as a basis, presented an annual budget to the Chinese Council for United States Aid (or its successor CIECD) and the USAID Mission to China. The annual program was derived from proposals submitted to the Joint Commission by its technical divisions, but the Commission itself made the final decision on program content.

Most individual projects supported by JCRR are proposed by sponsoring agencies and selected on the basis of their expected contributions to the production goals of the Four-Year Economic Development Plans.

Simplifying Procedures

In support of the changes made in its program objectives in 1963, mentioned in Chapter 2, and to simplify its supervisory and administrative procedures, JCRR has adopted the following operational guidelines:

1. The number of projects developed and executed shall be greatly reduced from the number prepared and approved in FY1963.

2. The number of project sponsors shall be similarly reduced, with the relevant Taiwan provincial departments and bureaus as the primary sponsors of JCRR-assisted projects.

3. Projects relating to established activities are to be de-emphasized and innovative and pioneering types of agricultural development efforts are to be stressed. Projects other than those of the innovative and pioneering types are to be transferred to the appropriate government agencies and private organizations as rapidly as possible.

4. Ongoing project activities that are given further assistance by JCRR in FY1964-FY1965 shall require at least one-third matching support from sponsoring agencies, and this ratio shall be further increased in succeeding years.

5. Much greater emphasis will be placed upon educational demon-

strations relating to research to assure that all available research results will be applied to agricultural production and marketing operations.

6. Projects involving additional buildings and other capital facilities and equipment will be given low priority in favor of projects which fully exploit the use of current capital investments.

7. Grants will be extended to revenue-bearing enterprises only when such enterprises serve as *bona fide* educational demonstrations for farmers or other agricultural entrepreneurs, or when the enterprise involves risks due to the introduction of innovations. Otherwise, only loans will be extended to revenue-producing organizations and agencies.

5. Budget Analysis

Rampant inflation multiplied JCRR's difficulties in financing its projects. Fortunately, the situation improved after the currency reform in June, 1949, in which the New Taiwan dollar replaced the Old. With an official conversion rate of NT$5 to US$1, there was only a mild inflation in the first few years of the currency reform. Under such circumstances, JCRR could begin to plan and implement its program on a larger scale.

United States economic aid in United States dollars to the Republic of China in the fifteen years from 1951 to 1965, inclusive, amounted to US$1,492.4 million, as shown in Table 5-1. Of the US$101.3 million devoted to the development of agriculture and natural resources, which accounted for 6.8 per cent of the total, US$10,629,550, or 10.5 per cent, was channeled through JCRR. In other words, JCRR has handled only 0.71 per cent of the total United States dollar aid to China.

The sale of United States aid commodities for use in helping to develop the eleven categories of activity listed in Table 5-1 totaled NT$25,530.3 million (equivalent to US$1,002.3 million), as shown in Table 5-2. Of the NT$4,833.1 million devoted to the development of agriculture and natural resources, which accounted for 18.9 per cent of the total, NT$4,086.02 million, or 85 per cent, was channeled through JCRR.

A breakdown of the US$ and NT$ United States aid funds funneled through JCRR from FY1951 through FY1965 is shown in Tables 5-3 and 5-4, respectively.

The two million odd United States dollars expended on United States advisers were used for the payment of salaries and the defray-

Table 5-1. Breakdown of total United States economic aid in US$ to the Republic of China, FY1951–FY1965

Category	Amount
Industry and mining*	US$ 673.8 M
Agriculture and natural resources†	101.3 M
Transportation	52.0 M
Health and sanitation	33.0 M
Military budget support	16.6 M
Education	9.3 M
Public administration	2.9 M
General and miscellaneous	19.1 M
Agricultural consumable commodities under PL480‡	306.1 M
Agricultural consumable commodities under supporting assistance‡	232.1 M
Nonagricultural consumable commodities‡	46.2 M
Total	US$1,492.4 M§

* This item includes US$159.6 million appropriated for the development of electric power.

† This item includes all forms of assistance—commodity, personnel, and TA training —as well as certain large-scale multi-purpose projects in which agriculture formed an integral part, such as the Shihmen Dam, for which US$23 million under this item was appropriated.

‡ These three items refer to AID-financed commodities in consumable form. The commodity components of the other categories in this table were industrial equipment, industrial raw materials, research facilities, etc.

§ Of the grand total of U.S. economic aid, US$1,270 million was in the form of grants and the remainder in loans.

Table 5-2. Breakdown of total United States economic aid in NT$ to the Republic of China, FY1951–FY1965

Category	Amount
Military budget support	NT$10,954.6 M
Agriculture and natural resources*	4,833.1 M
General and miscellaneous	3,307.3 M
Industry and mining	3,270.0 M
Transportation	1,637.1 M
Health and sanitation	756.9 M
Education†	698.1 M
Public administration	73.2 M
Total	NT$25,530.3 M

* This category includes all expenditures and funds programmed and spent by JCRR and also by the AID Mission to the Republic of China.

† A part of the funds spent on education was for agricultural education through the vocational agricultural schools and the university colleges of agriculture.

Table 5-3. U.S. dollar economic assistance to agriculture in the Republic of China channeled through JCRR, FY1951–FY1965

Category	Amount
U.S. agricultural advisers	US$ 2,070,000
Commodities	7,106,400
Participants training:	
in U.S.	1,033,850
in other countries	419,300
Total	US$10,629,550

Source: Office of U.S. AID Representative in China, November 1967.

Table 5-4. New Taiwan dollar economic assistance to agriculture in the Republic of China channeled through JCRR, FY1951–FY1965

Category	Amount
Project costs*	NT$4,025,113,000
Participants training local currency costs	23,050,000
U.S. technicians local currency costs	37,855,000
Total	NT$4,086,018,000

Source: Office of U.S. AID Representative in China, November 1967.
* This figure includes NT$20,008,000 programmed in FY1950.

ment of supporting US$ costs of 118 technicians, who provided a total of 189 man-years and eight man-months of service. The fields in which assistance was rendered included land and water resources development, crop and livestock production, research, extension work, the improvement of farm credit and cooperative facilities and systems, economic planning, forestry, rural health, and agricultural administration.

The United States commodities for whose procurement a total of US$7,106,400 was expended included such items as scientific research equipment, agricultural machinery and vehicles, plant protection equipment, warehousing and grain storage machinery, agricultural processing equipment, improved seeds and livestock breeding stock, visual aid materials and equipment, drugs and medical supplies.

Under the training program, which cost US$1,453,150, altogether 389 Chinese technicians were sent to the United States; 377 Chinese agriculturists, irrigation engineers, foresters, fishery experts, and rural health specialists were sent to Japan and other third countries for training. The training of specialists proved one of the best investments made for the development of agriculture in Taiwan.

Included in the NT$4,025 million expended on JCRR projects are NT$140 million used for the payment of salaries and supporting costs of Chinese agricultural technicians serving on the JCRR staff. This sum represents about 3.5 per cent of the project costs.

Allocation of Money for Major Categories

The JCRR allocation of money in NT$ and US$ for major project categories from FY1950 through FY1965 is shown in Table 5-5.

Table 5-5. JCRR allocation of money in NT$ and US$ for major project categories, FY1950–FY1965 (in thousands of dollars)*

Activity	NT$	US$
Crop production	415,581	1,024.4
Livestock production	293,237	348.9
Water use and control	1,241,378	2,552.4
Forestry and soil conservation	213,310	475.9
Rural organization and agri. extension	286,610	88.0
Economic research and agri. credit	376,654	11.4
Fisheries	307,426	30.5
Land reform	25,812	—
Rural health	142,375	317.5
Agricultural research and education	90,627	731.2
Rural electrification and communication	53,763	—
Government budget support to local agri. programs	242,946	1,027.0
Miscellaneous projects	150,485	194.5
Administration	184,909	304.7
Total	4,025,113	7,106.4

Note: The table includes only funds disbursed by JCRR. Salaries, allowances, and international travel expenses of American personnel, and expenses for sending trainees to the United States or other countries are paid by the AID China Mission. The US$ figures represent costs of commodity.

* Condensed from Appendix 1, JCRR *General Report XVII*, 1966, where the year-by-year figures are given.

Water use and control quite naturally accounted for the largest share (NT$1,241 million or 30.8 per cent, and US$2,552,400 or 36 per cent) of the total JCRR funds; the development of water resources and the construction of flood control works are long-term projects and call for larger investments than any other rural development projects. Two major categories, water use and control and crop production, accounted between them for over one half of the United States dollar aid, while over one half of the local currency went into these two and a third category, economic research and agricultural credit.

It is significant also that less money from United States aid (less than two-thirds of 1 per cent) was spent on the land reform program than on any other category of JCRR projects. Moreover, no US dollar aid was required either for land reform or for rural electrification and the development of rural communications.

Although agricultural research and education constitute one of the best investments in the agricultural sector, the work in this category received only 2.25 per cent of NT dollars and 10.29 per cent of US dollars channeled through JCRR.

Viewed as a whole, the JCRR funds (NT$4,025 million and US$7,106,400) expended in the sixteen years FY1950–FY1965 are less than the total export value (US$127,389,000) of two new crops, mushrooms and asparagus, for the three-year period, 1965–1967.

In addition to contributing these monetary returns, the development of agriculture resulting from JCRR operations has also stimulated industrial development in Taiwan. Thus, investment in agriculture, if it is associated with technological improvement, can be one of the best investments for developing countries.

Grants and Loans

Of the NT$4,025 million spent on JCRR projects from 1950 through 1965, NT$2,601 million were in the form of grants-in-aid and NT$1,424 million in the form of loans in support of revenue-producing enterprises at interest rates ranging from 6 to 12 percent per annum. The allocation of money for major categories of JCRR projects, showing both grants and loans, is shown in Table 5-6.

Water use and control received a larger share in grants (25 per cent) and loans (41 per cent), as it did in terms of NT$ and US$, than any other JCRR project category. Grants were made for flood control, planning, and surveys; loans for irrigation construction. In the western parts of Taiwan, where climate favors agriculture, 50 per

Table 5-6. JCRR allocation of money in grants and loans for major project
categories, FY1950–FY1965 (in thousands of NT dollars)

Activity	Grant	%	Loan	%
Crop production	386,437	14.86	29,144	2.05
Livestock production	177,734	6.83	115,503	8.11
Water use and control	662,147	25.45	579,231	40.68
Forestry and soil conservation	185,780	7.14	27,530	1.93
Rural organization and agri. extension	228,206	8.77	58,404	4.10
Economic research and agri. credit*	43,343	1.67	333,311	23.41
Fisheries	84,469	3.25	222,957	15.66
Land reform	25,812	0.99	—	—
Rural health	142,107	5.46	268	0.02
Agricultural research and education	90,397	3.47	230	0.02
Rural electrification and communication	15,499	0.60	38,264	2.69
Government budget support to local agri. programs	242,946	9.34	—	—
Miscellaneous projects	131,596	5.06	18,889	1.33
Administration	184,909	7.11	—	—
Total	2,601,382	100.00	1,423,731	100.00

* Though economic research and agricultural credit are put under one category in
Tables 5-5 and 5-6, the reader will please note that no loans have ever been required for
economic research; all the loans in this category refer to loans for agricultural credit.

cent of the irrigation construction cost was provided by the Taiwan
Provincial Government in the form of grants and 50 per cent by
JCRR in loans. In the eastern parts where the climate is unfavorable,
the Provincial Government provided 75 per cent in grants, and JCRR
25 per cent in loans. The record of loan repayment was excellent.
The farmers, whose crops were increased by the irrigation water thus
made available, repaid over 99 per cent of the loans as they fell due.

Next to water use and control comes crop production, which accounted for about 15 per cent of the grants. These two categories and
two others—government budget support to local agricultural programs, and rural organization and agricultural extension—took up
58 per cent of the total. The remaining 42 per cent was distributed,
in a descending order, among forestry, administration, livestock

production, rural health, outlying islands and mountain resources development, and other lesser items.

In the percentage of money received in loans, agricultural credit (23 per cent) and fisheries (over 15 per cent) ranked next to water use and control (see footnote in Table 5-6). These three items together absorbed almost four-fifths (79.7 per cent) of all the loan funds. One feature is noteworthy; no loans whatever were extended either to land reform or to government budget support to local agricultural programs. It may also be mentioned that no loans were ever required for administrative purposes.

Changing Pattern of United States Aid

The United States economic aid to China for the development of agriculture in Taiwan has never remained static but has changed with circumstances. This fluctuation is reflected in JCRR's annual project disbursements in local currency and in United States dollars.

The JCRR program in Taiwan began with an initial NT$20 million in FY1950; the budget gradually grew to NT$199 million in FY1956—an almost tenfold increase in six years. But the record high of NT$496 million was not attained until FY1962. (The JCRR budget of NT$534 million for FY1960 was an exceptional case, because over half was used to repair the unprecedented flood damage to crops and farm lands during 1959 and 1960.) In FY1965, the last year of United States aid to China, a respectable NT$357 million was still called for to maintain the momentum of agricultural development generated by the three successive Four-Year Economic Plans of 1953 to 1964.

Table 5-7 shows the number of JCRR projects and actual amounts spent from FY1950 through FY1965.

JCRR's disbursement of United States dollar funds as shown in Table 5-8 started with a meager US$63,000 in FY1952 and ended with a still smaller amount of US$55,600 in FY1962. In between these years there were ups and downs, reaching a peak of US$1,464,600 in FY1955, after which the one million mark was exceeded only in FY1959, when there was another upward swing to US$1,206,500.

Throughout the fifties an overwhelming percentage of the United States NT$ generated for agricultural aid in Taiwan had been in

Table 5-7. Number of JCRR projects and actual amounts spent, FY1950–
FY1965 (in millions of NT dollars)

Fiscal year	No. of projects	Actual amount spent	Increase or decrease
1950	115	20.0	—
1951	155	39.9	+19.9
1952	229	57.2	+17.3
1953	275	81.5	+24.3
1954	308	92.3	+10.8
1955	352	147.2	+54.9
1956	534	199.2	+52.0
1957	484	222.9	+23.7
1958	461	213.5	−9.4
1959	425	252.0	+38.5
1960	500	534.6	+282.6
1961	585	445.6	−89.0
1962	691	496.1	+50.5
1963	613	442.4	−53.7
1964	283	424.0	−18.4
1965	270	356.7	−67.3
Total	6,280	4,025.1	

Table 5-8. JCRR allocation of money in US$, FY1950–FY1965 (in thousands of dollars)

Fiscal year	Amount
1950	—
1951	—
1952	63.0
1953	382.1
1954	177.8
1955	1,464.6
1956	500.7
1957	651.4
1958	858.8
1959	1,206.5
1960	877.2
1961	868.7
1962	55.6
1963	—
1964	—
1965	—
Total	7,106.4

the form of grants and the rest in the form of loans. But in the first four years of the present decade, the percentage of grants gradually dropped.

Planning of JCRR Programs since 1966

Since the termination of United States economic aid to China in June, 1965, JCRR activities have been financed mainly with appropriations from the Sino-American Fund for Economic and Social Development (SAFED), which will be discussed in Chapter 23. The expenditures approved by the Joint Commission were NT$359 million in FY1966, NT$404 million in FY1967, and NT$446 million in FY1968; the allocations for the three-year period are shown in Table 5-9. These sums represented, for each of the three years, 17 per cent of the funds from SAFED for economic and social development in that year.

In addition to the appropriations from SAFED, two special funds from other sources were made available to JCRR to implement the programs for unified agricultural credit and family planning, which will be separately treated in Chapters 19 and 20. From FY1966 through FY1968, the funds provided for the former were NT$54 million and for the latter NT$44 million.

Of the total NT$1,209 million made available to JCRR from SAFED during these three years, NT$612 million were in the form of grants and NT$597 million in the form of loans. The largest share of grants-in-aid went to water use and control (23 per cent), followed in a descending order by crop production (17 per cent), administration (15 per cent), forestry and soil conservation (11 per cent), and other categories (less than 10 per cent each). The percentage for rural health was small because family planning did not require financing from SAFED. The 15 per cent for administration, which is much higher than the 7 per cent for the same purposes during FY1950–FY1965, as shown in Table 5-6, is to be explained by the fact that total grants diminished (NT$373 million in FY1960 versus NT$202 million in FY1968) whereas the JCRR staff was kept at about the same size, since it is the minimum required for the efficient administration of a technical organization of its kind.

A notable change took place in the distribution of loans among the

Table 5-9. JCRR budget allocation from Sino-American Fund for Economic and Social Development, FY1966–FY1968 (in thousands of NT dollars)

Item	Grant	Loan
Crop production	104,190	14,000
Livestock production	39,220	177,740
Water use and control	139,225	135,967
Forestry and soil conservation	68,782	42,750
Research and training in agricultural activities	46,964	—
Farmers' service*	38,895	49,752
Fisheries	23,770	155,778
Rural health†	18,120	—
Other supporting programs	16,737	—
Outlying islands	25,778	21,259
Administration	90,142	—
Total	611,823	597,246

* Not including funds for agricultural credit.
† Not including grants allocated from special fund for family planning.

various project categories during the three fiscal years of 1966, 1967, and 1968. Livestock took the lead, receiving 30 per cent of the total loans, followed by fisheries with 26 per cent, and water use and control with 23 per cent. The sudden emergence of livestock production to prominence was due to the expansion of the hog-raising program by the Taiwan Sugar Corporation, which financed it with NT$32 million in FY1967 and NT$48 million in FY1968. The expansion of tuna and shrimp fishing called for more funds for fisheries, which, therefore, forged ahead to second place. Water use and control dropped to third place in so far as loans were concerned, because projects in this category received some of the money they needed from the newly established Joint Irrigation Fund and the Flood Control Fund, described in Chapter 12.

The interest rates charged on loans for various kinds of projects since July 1, 1965, have been: 10.08 per cent per annum for revenue-producing projects of an ordinary nature; 8.28 per cent per annum for projects that have passed the preliminary demonstration and experimentation stages, whose profits are still uncertain; 6 per cent per annum for the rehabilitation of damages done by natural disasters, for new or experimental projects or innovative-type projects that cannot make any profit during the initial stage, for the construction of nonprofit public facilities, and for projects formerly financed by grants.

The different interest rates charged on loans for different kinds of projects are another instance of how JCRR budgets and programs have remained dynamic and flexible to meet the needs of changed and changing times.

PART II. GROUNDWORK IN TAIWAN

6. Initial Years in Taiwan

Lying 100 miles off the southeastern coast of the Chinese mainland and with the Tropic of Cancer cutting across it just in the middle, the island of Taiwan has a subtropical climate in its northern half and a characteristically tropical one in the southern. The Central Mountain Range, which runs from north to south, covers about two-thirds of the 35,960 sq. km. (13,836 square miles) of the land area. The annual rainfall in the important agricultural regions averages 1,763 to 3,042 mm (71 to 122 inches). Like the Philippines to the south, Taiwan is under the constant threat of typhoons that originate in the southern Pacific in the summer months and early fall; since they are particularly frequent from June to September, when most crops are growing, they have done much damage to both farm land and crops. On the average, three or four such tropical storms, with heavy winds and torrential rain, have swept over the island every year since 1897.

Because its climate is conducive to the decomposition of organic matter and the leaching of nutrients, Taiwan's agricultural land is relatively infertile. The tilth and aeration of the soils on the only sizeable plains in the southeastern part of the island are generally poor, because the soils are mostly of silty clay texture with massive subsoil structure and high bulk density. To make matters worse, torrential tropical rains may at any time carry away the surface soil from the hillsides. Therefore, to support a year-round cropping system, the soil requires, among other things, ample amounts of compost and fertilizers.

On the other hand, the different altitudes from the mountains down to the plains below provide a variety of climates—cold, temperate, subtropical, and tropical—in which different crops can be grown. They offer many opportunities but also demand a high level of cultural technology and intensive farming operations on the part of the farmers.

In the fifty years of Japanese occupation, from 1895 to 1945, Taiwan was a mere colony, exporting food to Japan to help feed her people, and serving as a market for her industrial products. During this period farm production was more than doubled by agricultural research and demonstrations, extension of *japonica* rice varieties, development of irrigation facilities, and use of chemical fertilizers. Farmers' associations were organized mainly for the collection of rice and other farm products for export to Japan and also to provide agricultural extension services.

However, as a result of deteriorated irrigation facilities, shortage of chemical fertilizers, and other adverse conditions during World War II, food and other agricultural production declined sharply to the level of 1910.

At the end of World War II in 1945, the Chinese National Government took over the administration of Taiwan, which once more became a Chinese province. A program for the rehabilitation of agriculture and industry was soon launched. For this purpose, large groups of administrators, scientists, agriculturists, and engineers were sent from the mainland. They, together with those originally in Taiwan, worked hard and with remarkable success at reconstructing the shattered economy of the island. Industrial plants destroyed by Allied bombing or fallen into disrepair were restored and made to run again. The irrigation systems and flood control works were repaired one after another. Under these measures, agricultural production showed steady improvement from 1946 to 1948.

The fall of the Chinese mainland to the Communists in December, 1949, forced the National Government to transfer its seat to Taipei, which thus became Free China's capital city. The ensuing period was a critical one in the development of Taiwan. Its population, suddenly swollen by wave after wave of civilian refugees fleeing the mainland, rose from 6.8 million in 1948 to 7.6 million in 1950, not including the 600,000 men in the armed forces, who withdrew to Taiwan on governmental orders to carry on the war against the newly set-up regime at Peiping. To keep the people adequately fed, special emphasis was laid on boosting food production.

JCRR's Transfer to Taiwan

JCRR was forced to transfer its headquarters from Canton to Taiwan in August, 1949. There it found only the vestiges of many agricultural agencies, all poorly maintained for lack of qualified personnel and funds. The farmers' associations had to be completely reorganized and overhauled. The extensive damage done by war to the farm lands and other agricultural assets had to be repaired before more production could be expected.

Fortunately, JCRR enjoyed one clear advantage: the inhabitants in Taiwan, being descendants of immigrants from the mainland or recent immigrants themselves, have all retained the cultural background and ways of life of their motherland and are no whit different from their brothers and sisters on the other side of the Taiwan Straits. Such being the case, the principles and procedures that JCRR had formulated and tried out in the mainland provinces (referred to in Chapter 2) proved equally workable in Taiwan and required only slight readjustments or modifications to suit the local conditions.

After sizing up the situation, the Joint Commission adopted a set of broad principles to guide its operations on Taiwan. First of all, it decided to give careful consideration to the amount of program funds that could be safely expended without exerting any adverse effect on the currency stabilization program of the Taiwan Provincial Government then in progress, and to limit its program accordingly. Second, it decided to develop the program already under way before considering any new projects. Third, it decided to limit new projects in general to three types: (1) those which would contribute to an immediate increase in agricultural production; (2) those which would meet other basic needs of the rural people, including the aborigines, such as farm rent reduction and rural health improvement; and (3) those of an investigatory nature, in fields directly related to agricultural production and improvement of the living conditions of the rural people.

At this time, a Food and Fertilizer Distribution Division was created to take over the functions of the Taiwan Regional Office, which had been abolished along with all the other Regional Offices on the mainland. The Agricultural Improvement Division, which became the Plant Industry Division after its functions relating to animal industry, farmers' organization, and rural economics were assigned to other newly created divisions, bore a greater share of the responsibilities for projects than any other division. Indeed, it served as a trail blazer for most of the JCRR activities in Taiwan.

Important Projects

A few important projects undertaken during 1949–1951 may be cited to show the JCRR approach in the initial years of its operation in Taiwan.

The JCRR agricultural improvement program covered many projects: (1) breeding of rice, wheat, sugar cane, pineapples, oranges, and vegetables, increasing the seed supply, and extending the acreages; (2) pest control of these crops; and (3) demonstrating and promoting the use of chemical fertilizers, green manure, compost manure, and the repair of farmers' compost houses.

Since this program consisted of so many diverse projects, it was criticized at the time as being rather piecemeal in nature. However, many of the projects were interrelated or mutually complementary; carried out simultaneously, they in fact led to the production of more and better crops. As a result of increased rice production, Taiwan became self-sufficient in food supplies and eventually had a surplus for export.

Another aspect of the JCRR program in the 1949–1951 period had to do with reforestation and the planting of windbreaks. Due to deforestation both during and after World War II as well as the illegal cultivation of hilly slopes for planting bananas, serious soil erosion had taken place on the hillsides. Many windbreak forests planted along the seacoast before World War II for the protection of farm crops against the northeast monsoon had been cut down by ignorant people for fuel. To cope with the situation thus created, JCRR assisted governmental and private agencies for reforestation and the planting of windbreaks in large areas. In addition, JCRR conducted a comprehensive survey of the main forest districts and the principal logging stations in 1951, and published a report entitled

Forest Conditions in Taiwan that served as a blueprint for the development of forestry in Taiwan in later years.

Other rehabilitation and development measures taken during this period included those in irrigation, animal industry, and fisheries. Also initiated were programs for land reform, the reorganization of farmers' associations, rural health, and education and training. All these undertakings will be separately treated in the following chapters.

7. Land Reform

All Chinese peasants on Taiwan, like their brothers on the mainland, have a deep urge to own the land they till. But formerly most of the farm lands on the island were the estates of big landlords and not for sale. Any land available for purchase was priced so high that no peasant could afford it. The competition for land was keenest in the densely populated areas where the soil was fertile and irrigation facilities good. The majority of the farming population were tenants who paid high rent for the land they had leased. Farming was on a very small scale and farm incomes were low. However hard a peasant might work, he could seldom save enough to buy land. The only way he might acquire it was by inheritance from some childless relative. The climb up the rural social ladder—farm hand to tenant to owner-operator—was very difficult, if not impossible.

In 1949 the farming population in Taiwan constituted a little more than half the entire population and included the following categories: owner-operators, 34 per cent; part-owners who cultivated both their own land and some leased from others, 23 per cent; full tenants, 36 per cent; and farm hands, 7 per cent. Tenant farms occupied about 41 per cent of the cultivated area. The average size of a farm operated by a tenant family was about one hectare. In the pre-reform days, the tenant generally had to pay to the landlord one half or more of the yield from the main crop—that is, rice for paddy land and sweet potatoes for upland. The tenant farmer, weighed down by excessive rentals, was poor and miserable.

As to the landlords, the life of ease and leisure they led was made possible by the toils of their tenants. Hence no love was lost between the two parties, and there was latent unrest in the rural districts,

56

making them susceptible to Communist propaganda. Capitalizing on the situation, the Chinese Communists posed as "agrarian reformers" and champions of the underdog. By 1948, while conditions on the mainland were fast deteriorating, the villages in Taiwan, which had been newly recovered after fifty years of Japanese colonial rule, were also showing signs of unrest and instability. There was real apprehension that Mao Tse-tung, after gaining control of the mainland, might push on to attack this island province as well. To forestall any such possibility, the Taiwan Provincial Government decided to improve the lot of the peasants by introducing the first phase of a sweeping land reform program in the form of farm rent reduction.

Farm Rent Reduction

JCRR assisted the Chinese Government's efforts to carry out land reform by helping to plan, train personnel, and conduct cadastral surveys in the initial phase of farm rent reduction.

The Taiwan Provincial Government initiated a general land reform program by promulgating in April, 1949, a set of regulations governing the lease of private farm lands in Taiwan. According to these regulations, no farm rent could exceed 37.5 per cent of the total annual yield of the main crop (rice or sweet potatoes). Any rent in excess of this amount should be reduced; any that was less should remain unchanged. In addition, the verbal contracts that had been customary would be replaced by written leases; these would specify all the terms, would be signed by the parties concerned, and duly registered with the local authorities. As the program gathered momentum, another law, officially known as the Farm Rent Reduction to 37.5 Per Cent Act, was passed by the national legislative body and promulgated by Presidential decree in 1951. This basic law and a few other supplementary rules and regulations issued by the Taiwan Provincial Government completed the legal framework within which the farm rent reduction policy was carried out.

Before rents could be adjusted, a standard yield had to be fixed for each of the twenty-six grades of farm land into which all arable land in Taiwan is divided on the basis of its soil fertility and productivity. The work of appraising the standard yields of the different grades of paddy fields and dry land was initially done by local farm tenancy committees, which submitted their findings to county or city farm tenancy committees. The decisions of these committees were subject to approval by the Provincial Government.

In addition to farm rent reduction, the program also provided the security of tenancy rights in several ways. First, the tenure of the farm lease was extended to a minimum of six years and could be renewed at the request of the tenant. The rights of the tenant were fully guaranteed, and he could not be arbitrarily evicted, as he had in the past. Only if the tenant died without leaving an heir, or waived his right of cultivation by migrating or changing his occupation, or fell behind in his stipulated farm rent by a cumulative total of two full years' payment could the farm lease contract be terminated by the landlord before the date of its expiration.

Second, the landlord was forbidden to collect the farm rent in advance or demand any security deposit. Whatever security deposit already paid must be refunded.

Third, the amount, kind, quality, and standard of the stipulated farm rent, the date and place of payment, and other relevant matters were also clearly stated in the lease contract. If the rent payable in kind was to be delievered by the tenant to the landlord, the latter must pay the cost of delivery according to the distance covered.

Finally, to be legally binding, all new lease contracts and the revision or termination of existing ones had to be formally registered with the proper authorities, who saw to it that all the provisions were strictly enforced.

To help implement the rent reduction program, local farm tenancy committees, mentioned earlier, were organized. Each committee consisted of eleven members, five representing tenant farmers, two representing owner-operators, two representing landlords, and two appointed by the local government. The committees were charged with the important functions of appraising the standard yields of the total annual main crop, investigating crop failures caused by natural disasters and recommending measures for the partial reduction or total remission of rents, and conciliating disputes over leases between landlords and tenants.

The effects of farm rent reduction were immediate. They were reflected in improvements in the farmers' livelihood, increases in agricultural production, elevation of the tenant farmers' position, and an appreciable decline in the price of farm lands.

Improvements in the livelihood of tenant farmers were manifested in more food to eat and better nutrition, in the new clothes they and their dependents wore, in repairs to the formerly dilapidated farmhouses, and in the daily necessities they now could afford to buy with the proceeds from the sale of surplus products.

With extra money from lower rents, tenant farmers were able to buy farm equipment and fertilizers; thus agricultural production was increased. In the pre-rent reduction days, no tenant farmer could afford to own a buffalo, and two or three tenant farmers usually shared the ownership of one buffalo. But now each could buy a draft animal for his exclusive use.

In the old days the tenant was the social inferior of his landlord, who could oppress and exploit him in more ways than one. Now that the signing of the new lease contract placed them on an equal footing, the tenant-farmer could enjoy the full protection of the law, and his social status was elevated.

The reduction of farm rent made the landlords less eager to own land. Some of them sold and invested in more lucrative businesses. This caused a decline in the price of farm land and enabled the tenant farmers to buy it with the increased income they now received as a result of lowered farm rent and increased production. According to a survey conducted by the Taiwan Provincial Land Bureau, 61,000 tenants purchased 32,000 hectares of farm land in the five-year period 1949 to 1953.

Sale of Public Lands

Having laid a solid foundation for land reform with the successful implementation of farm rent reduction, the Chinese Government proceeded to sell by installments the farm lands it had acquired following the liquidation of the Japanese colonial administration at the end of World War II. The Government did this partly to enable more tenant farmers to become owner-operators and partly to set an example for the landlords who would, during the third stage of the land reform program, be required to sell their lands to their tenants.

These public lands taken over from the government and nationals of Japan in 1945 comprised approximately 180,000 hectares—nearly one fifth of the arable land in Taiwan. From 1951 to 1961, some 96,000 hectares were sold to 156,000 farm families which paid for their purchases 330,000 metric tons of paddy rice and 882,000 metric tons of sweet potatoes. The revenue thus realized by the Government was used primarily in the implementation of other land reform measures.

Most purchasers of the public lands were tenant farmers who badly needed land of their own. In principle, the Government set the price at a reasonable level, not to exceed the current market price. Specifically, it was calculated in terms of farm products at 2.5 times the value of the annual main crop, payable in kind in semiannual installments over a period of ten years. Each purchaser family was entitled to buy a maximum of one-half to two hectares of paddy land or double the amount of dry land. But the incumbent tiller of public land lawfully leased to him by the Government might be allowed to purchase all the land originally leased to him without being subject to this restriction.

In buying the public lands, the farmer purchasers actually did not have to pay more than what they would have had to pay in annual rents had they remained mere tenants. Under this new arrangement they became owners of the lands they were tilling after the payment of the last semi-annual installment. Moreover, the purchasers were given the right to shorten the period of payment of the purchase price if they wished. This was another provision for their benefit.

It must be pointed out that not all the public lands were offered for sale to the farmers. As of 1969, some 53,000 hectares were retained for use by public bodies such as the Taiwan Sugar Corporation (a provincial enterprise), schools, and military establishments. But the sale of part of the public domain to the incumbent tillers and others needing land was a manifestation of the Government's determination to go through with the land reform program.

The Land-to-the-Tiller Act

The third and last phase of the land reform carried out in Taiwan is known as the Land-to-the-Tiller Program, by which is meant the compulsory purchase by the Government of the landlord's tenanted farm lands in excess of a certain limit and their resale at the same price to the incumbent tillers.

But before such an ambitious project could be launched, a general landownership classification was necessary. Since previously the cadastral records had been compiled according to land plots, one could tell which particular plot belonged to whom, but not how much land landlord "X," "Y," or "Z" owned throughout the province. To supply this deficiency, new land record cards and landownership cards were compiled from the information contained in the old land registers. On-the-spot checks were made to clear up doubtful points.

This spade-work, done in 1951 and 1952, yielded some six million cards showing all details about the location, area, grade, land category, owner's name, and actual user of each plot of land as well as the total landholdings of each individual. On the basis of this information, the Taiwan Provincial Government drew up a preliminary draft of the Land-to-the-Tiller legislation for submission to the Central Government. This bill, as introduced by the Cabinet with certain minor changes, was finally passed by the national legislature on January 20, 1953, and promulgated by Presidential decree six days later.

The principal provisions of the Land-to-the-Tiller Act may be summarized as follows:

As the Land-to-the-Tiller Act applied only to farm lands under lease to tenants or cultivated by farm hands, a landowner who cultivated his own lands with the help of his dependents would not come under its provisions. Even a landlord who put his lands to lease or employed farm hands to do the tilling for him was allowed to retain for himself a maximum of three hectares of medium-grade paddy land or six hectares of medium-grade dry land, or equivalent amounts of paddy and dry land of superior or inferior qualities, as the case may be. All tenanted lands in excess of this limit were subject to compulsory purchase by the Government for resale to the incumbent tillers, who might be either tenants or farm hands. But religious institutions and family clans with landed property devoted to ancestral-worship purposes were allowed to keep twice the amount of tenanted lands that individual landlords might retain.

The price of farm lands compulsorily purchased from landlords was 2.5 times the value of the average annual main crop for the respective land grades. The landlords were paid for their lands 30 per cent in stock shares of four government enterprises and 70 per cent in land bonds redeemable in kind. The use of the stock shares of public enterprises in part payment of the land price was aimed at converting those enterprises into private ones and at the same time encouraging the landlords to invest their money in industry rather than in lands, as they had done for centuries in the past. The land bonds in kind were of two sorts—rice bonds and sweet potato bonds. Rice bonds upon maturity were redeemable in rice; sweet potato bonds were redeemable in cash calculated in terms of the prevailing market price of sweet potatoes at the time of redemption. All land bonds bore an annual interest of 4 per cent and were redeemable in twenty semiannual installments spread over a period of ten years.

These two methods of payment were designed to avoid causing an inflationary tendency by a sudden increase in the amount of currency in circulation. They were special devices worked out by the Government to meet the requirements of the time.

The farm lands thus purchased from the landlords were resold at cost by the Government to their incumbent tillers. The price of paddy lands was paid in rice, except in certain areas where cash was accepted as a substitute; the price of dry lands was paid in cash equivalent to the prevailing local market price of sweet potatoes at the time of payment. The farmer purchaser was required to pay the land price plus interest at 4 per cent per annum in twenty semiannual installments spread over a period of ten years. But the amount any farmer purchaser had to pay in a year, including both his semiannual installments and the land tax he owed as landowner, was never to exceed what he had formerly paid as a tenant.

The Land-to-the-Tiller Act also provided that whenever the landlord wanted to sell the farm land he was entitled to retain under the law, the incumbent tiller should have first priority of purchase at a price to be negotiated by the parties concerned or set by the local farm tenancy committee, referred to above, if the negotiations failed. The prospective buyer might also request the Government for loans for the purchase he intended to make.

To help improve land use and increase farm production, cheap loans from a special production fund established by the Government were available to farmer purchasers upon request.

If a force majeure damaged any farm land purchased under the Land-to-the-Tiller Program so severely that it became partly or wholly unusable, the farmer-purchaser might request the Government to grant him a certain reduction or exemption of the installment payments of the purchase price yet to be made. Similarly, in the event of serious disaster or crop failure, he might be allowed to postpone one or more of the semiannual installment payments due, with the proviso that he would later pay all the installments thus postponed.

The Land-to-the-Tiller Program was equitably conceived and warmly supported by public opinion. The landlords were satisfied with the reasonable compensation paid them for their lands. The tenants and farm hands welcomed the opportunity to become owners of the lands they had been tilling for many years and, in certain cases, even for generations.

Financing the Land Reform Program[1]

As mentioned earlier, the purchase of the lands by the Government for resale to tenants did not involve any monetary payment but was effected through two unique devices: the issuance of land bonds in kind and the transfer of the stock shares of four government-owned industries to the landlords as compensation for their lands. The only monetary outlay was for administrative expenditures; this totaled NT$265 million, of which JCRR contributed less than 10 per cent. The other agencies, which contributed over 90 per cent of the funds and the services of their personnel for the implementation of the various phases of the land reform program, were the Taiwan Provincial Land Bureau, the Taiwan Provincial Food Bureau, and the Land Bank of Taiwan. The administrative cost included the salaries and travel allowances of extra personnel employed by the three

[1] Chen Jen-lung, "The Administration of Agricultural Land Reform in Taiwan" (presented as a country report under the title "Land Reform in the Republic of China" by the Chinese delegation to the Second World Land Reform Conference sponsored by FAO/UN and held in Rome, June 20–July 3, 1966).

agencies and local governments, but did not include salaries paid to their regular staff members assigned to carry out certain aspects of the program.

On the other hand, the Government realized NT$660 million from reselling the landlords' lands to their tenants under the Land-to-the-Tiller Program. (This sum represented the cash value of stock shares of the four government-owned industries handed over to landlords in part payment for their lands under the Land-to-the-Tiller Program.) In addition, it realized NT$821 million from selling public lands to tenants—making a total of NT$1,481 million. Part of this sum was invested by the Government in irrigation works such as the Shihmen Dam, and the rest was used to establish a special fund for follow-up land reform measures such as land consolidation.

Results of Land Reform[2]

Studies by JCRR in 1957 and 1959 showed that significant changes followed the reform measures. To begin with, there has been a marked increase in the number of owner-operators, who now have greater incentives to work harder and produce more on their small family farms.

Carried out through peaceful and legal means the Land-to-the-Tiller Program, coupled with rent reduction, has enabled the farmers to obtain more income from farming. The share of farm income for labor wages has greatly increased. It went up approximately 9 per cent from 1950 to 1955, whereas farm rentals paid to landlords were cut by about 14 per cent.

Since farmers have had larger incomes, they have tended to spend more for personal consumption, particularly of nonagricultural products; thus their standards of living have risen. On the average, they spent 6 per cent more of their family income for nonagricultural products and 2 per cent less for agricultural products, thus making up 4 per cent more of their family incomes for total consumption in

[2] T. H. Shen, "Land Reform and Its Impact on Agricultural Development in Taiwan" (presented before the International Seminar on Land Taxation, Land Tenure, and Land Reform in Developing Countries, sponsored by the Lincoln Foundation, Phoenix, Arizona, U.S.A., held at Taipei, Dec. 11–19, 1967).

1955 than they did in 1950. As a result of a gain in the consumption ratio, their savings ratio dropped from 14 to 10 per cent in the same period.

Another change that has occurred as a result of land reform is in the investment pattern. The total investment in agriculture in 1955 was about twice the amount in 1950, while the net farm income registered a fourfold increase during this period. This means that the percentage of net farm income reinvested in agriculture was relatively lower in 1955 than in 1950. Farmers' savings accounted for 51 per cent of the sources of investment funds in 1950 but dropped to 32 per cent in 1955, owing partly to the fact that a portion of the savings had been diverted for use as installment payments on the land purchase. On the other hand, public investment in agriculture in the form of irrigation improvement and other measures, which accounted for only 18 per cent in 1950, rose to 25 per cent in 1955. From this it is clear that public investment has played a greater part in agricultural development since the implementation of land reform.

Along with changes in the investment pattern, there have also been certain changes in the production pattern. Total agricultural production increased by about 21 per cent from 1950 to 1955, including a 15 per cent hike in crops and 92 per cent in livestock and poultry. During this period the production of sugar cane remained fairly stable, but rice and other crops registered 14 and 23 per cent increases, respectively.

Equally interesting have been changes in agricultural inputs. While the crop area remained without much change in the 1950–1955 period, labor input increased by 8 per cent, input in fixed capital by 25 per cent, and working capital as measured in terms of constant New Taiwan dollars and spent mainly for the purchase of fertilizers, feeds, and pesticides, by as much as 75 per cent. This phenomenal increase in working capital constituted the most important single input leading to increased agricultural production. In other words, technical improvements in agricultural production were the result of biological innovations in crop and livestock production, which called for more working capital input.

The land reform program in Taiwan was introduced on the assumption that it would provide the farmer with greater incentives for production and have a favorable effect on the productivity of agricultural resources. This assumption has been borne out by subsequent developments. In the first five years from 1950 to 1955, the aggregate productivity of farm resources in Taiwan advanced by about 7 per cent, due mainly to the intensive use of land.

A small sample survey of landlords conducted in 1959 showed that about 64 per cent of the payments they received for the sale of their lands under the Land-to-the-Tiller Program they used for investment (42 per cent) and for personal consumption (22 per cent); the other 36 per cent they held in the form of land bonds pending maturity in later years. Since the landlords invested in industry and housing, it may be concluded that they have made some contribution to the industrial development of Taiwan since land reform.

Socially, land reform in Taiwan has provided security for tenant farmers, raised the social status of the rural people and enabled them to participate in rural organizations, including farmers' associations, farm irrigation associations, and the farm tenancy committees, which are all democratically operated for their benefit. At the same time, the people in rural areas have become more politically conscious, as evidenced by their greater interest in local elections. It may be concluded that the land reform program in Taiwan has provided a favorable climate for the betterment of the livelihood of the rural people and the elevation of their social status, in addition to laying a sound foundation for social advancement and agricultural development.

A survey of the economic conditions of landlords and tenants made by JCRR in cooperation with the Ministry of the Interior in 1969[3] showed that, under the present system of tenancy, the landlord obtains from each hectare of tenanted land about NT$6,548, out of which he has to pay NT$1,500 in land tax, and the tenant a net income of NT$14,747. Post-reform landlords depend very little for their livelihood on rent income, which constitutes only a tenth of their total family income anyway. But their standards of living are considerably higher than those of the tenants, owing largely to incomes

[3] Ministry of the Interior and JCRR, *A Survey of the Economic Conditions of Landlords and Tenant Farmers in Taiwan* (in Chinese) (Taipei: Ministry and JCRR, 1969), pp. 57–61.

from nonfarm sources. The average nonfarm receipts per household are NT$62,948 for the landlord and NT$17,181 for the tenant. Though landlords continue to exist, their economic influence has been much curtailed and they are no longer in a position to exploit the tenants. They have to make their own living, as the tenants do, by honorable means.

Impact of Land Reform on Industrial Development[4]

Under the age-old system of land tenure in Taiwan, as well as on the Chinese mainland, most farm lands were concentrated in the hands of wealthy landlords; the majority of the peasants were tenant farmers eking out a miserable existence. The villagers had little purchasing power, and there was little effective demand for industrial goods. Under these conditions, the climate was unfavorable to industrial investment. Only a thoroughgoing program of land reform, such as that carried out in Taiwan in the fifties, could stimulate industrial development.

In pre-reform days most industries in Taiwan were devoted to the processing of farm products. Other forms of industrial production offered little attraction to investment capital in view of the comparatively handsome returns on investment in farm lands. The landlords were more interested in keeping their landed property handed down from father to son than in selling it in exchange for industrial investment.

But as a result of the Land-to-the-Tiller Program implemented in 1953, the landlords were encouraged to invest in industries. Since part of the land price due the former landlords was paid in stock shares of four public enterprises, much capital that had been tied up in land for generations was released for industrial investment. Incidentally, this also served to give a big push to private enterprise.

The increased production and income obtained by farmers led to an increase in their purchasing power, which, in turn, gave rise to an expanded consumer market and stimulated industrial produc-

[4] C. F. Koo, "Land Reform and Its Impact on Industrial Development in Taiwan" (presented before the International Seminar on Land Taxation, Land Tenure, and Land Reform in Developing Countries, sponsored by the Lincoln Foundation, Phoenix, Arizona, U.S.A., held at Taipei, Dec. 11–19, 1967).

tion. At the same time, the farmers were now able to save part of their income, and their savings contributed a due share to domestic capital formation.

Farm production, besides providing for an increased domestic consumption, also provided a surplus for export. The foreign exchange earned by agricultural exports could be utilized for the procurement of capital equipment and raw materials for the further expansion of industry in Taiwan.

With a more than sufficient supply of food at home, the food price was stabilized. With the stabilization of the food price, other commodity prices and wages also remained relatively stable. With no violent fluctuations in the general price level, a favorable economic environment was created for industrial development.

Relevancy of the Taiwan Experience to Other Countries

One particularly interesting aspect of farming conditions in Taiwan should be noted in connection with the land reform that has been carried out there. The tenant farmer on this island usually manages his own farm with the help of his family. Such being the case, when the landownership was transferred under the Land-to-the-Tiller Program, the original tiller continued to cultivate the same piece of land and, of course, without any change in management. This procedure is much simpler than that in some other countries where the tenant supplies only labor and the landlord provides all capital and is also responsible for managing the farm himself.

The new owner-operators who have emerged following the land reform in Taiwan are industrious and intelligent. They not only paid all the installments of the purchase price when they fell due, but also have made and are making better use of land, labor, and capital. Being literate and well-informed, they are always ready to adopt new cultural techniques that increase farm productivity and income. Above all, they are business-minded and let themselves be guided by the profit motive, which is after all the best incentive to agricultural production.

Land reform alone is not enough. It must be combined with other measured: agricultural technology must be improved and farming operations bettered; there must be institutional changes such as the reorganization of farmers' associations; rural credit must be made available and a sound price policy established. There must also be follow-up work such as land consolidation. All these measures, which will be separately treated in later chapters, have made possible the degree of agricultural development that has been attained in Taiwan.

A crying need for land reform arises in many areas, and it becomes critical whenever the tenant farmers are too discouraged by the existing land tenure system to adopt new technology for production, because they have no assurance under the system that a reasonable proportion of the gains to be had from its adoption would accrue to them. It seems that such a need still exists in the developing countries. But just how each country should reform its system of land tenure must be considered in the light of its physical and economic environment and historical and cultural background. No standard formula can be applied; measures that have worked successfully in Taiwan may or may not work with equal success in other countries. What has been done on the island is only one example of how land reform can be carried out in a society that wants to free itself from the bonds of the past.

8. Farmers' Associations[1]

From the very beginning JCRR has recognized that well-organized farmers' associations are essential instruments for carrying out rural reconstruction programs.[2] Through twenty years of experience, it has found that projects for increasing agricultural production and improving the general well-being of rural communities can be successfully executed only by the farmers' own efforts, and has formulated its policies and programs accordingly.

There are a number of farmers' organizations in Taiwan, but the largest and most outstanding are the farmers' associations. Though JCRR has paid a good deal of attention to strengthening and improving the farm irrigation associations, the fishermen's associations, and specialized agricultural cooperatives as well, the major emphasis has been placed on the farmers' associations because of the multi-faceted services they render to their members and to the government and the community at large. An understanding of the farmers' associations will give one an idea of how the rural community in Taiwan is organized for mutual benefit. The other rural organizations are generally patterned after the FAs.

The farmers' associations in Taiwan are a federated system of cooperative organizations which render credit, purchasing, marketing, and agricultural extension services to their members. The organizations are directly controlled and operated by the farmers for the promotion of their own welfare.

[1] T. H. Shen, "The Farmers' Association in Taiwan," *Agricultural and Land Programs in Free China* (Taipei: Government Information Bureau, 1954), pp. 13–23.
[2] JCRR, *General Report I* (Taipei: JCRR, 1950), pp. 91–92.

There are altogether 364 farmers' associations; 341 are at the township level, 22 at the county and city level, and one is at the top, at the provincial level. Membership in the grass-roots farmers' associations at the township level is limited to one individual from each farm family. The members of each village in a township organize into a small agricultural unit (SAU), of which there were 4,871 with a total enrollment of 878,651 in 1967. These SAUs are the foundation on which the township farmers' associations are built.

Organization and Reorganization

Each small agricultural unit has a chairman elected by its members who convenes and conducts all meetings of the unit and carries out its decisions. Three or four representatives are elected by each SAU of a given township to attend the annual convention of the township farmers' association. It is in the annual convention that a board of directors, usually composed of eleven to twenty-one members, and a board of supervisors composed of three to seven members are elected. The board of directors constitutes the governing body of the association. It elects its own chairmen and appoints a paid general manager to carry out its policies and decisions. The board of supervisors inspects and audits the accounts and exercises general supervision over the conduct of the affairs of the association.

Similarly, each township farmers' association in the same county elects two or three members to represent it at the annual convention of the county farmers' association, which is organized in much the same way as the township farmers' associations.

Likewise, each county farmers' association elects two or three members to represent it at the annual convention of the Provincial Farmers' Association, (PFA) which, in its turn, is organized somewhat like the lower echelon FAs. Thus, the system of farmers' associations in Taiwan has a broad base at the township level with thousands of small agricultural units as basic components and rises through county farmers' associations until the apex is reached at the provincial level. The whole structure may be compared to a pyramid with the Provincial Farmers' Association at the top.

The farmers' associations in Taiwan have had a long and checkered history[3] since the first one was organized in 1900. During the Japanese occupation of the island, the associations were under the direct control of the Japanese colonial office, and all their chairmen, vice-chairmen, directors, and counselors were either appointed or selected by the colonial administration. For instance, the Administrative Officer of the Japanese Governor-General's Office was concurrently chairman of the PFA and the local magistrates were concurrently chairmen of the local associations. In this way the Japanese rulers of Taiwan used the farmers' associations as instruments of colonial rule in so far as agricultural production was concerned.

As part of the Japanese colonial heritage, the farmers' associations were dominated by landlords and the local gentry who knew nothing about agriculture and farming and cared still less. Soon after the removal of JCRR headquarters from the mainland to Taiwan, the Joint Commission realized the need for a sweeping reform of the associations so that they might become a more effective system for the benefit of the rural community and might be democratically controlled by the farmers themselves. To make a study of the question, W. A. Anderson, professor of rural sociology at Cornell University, Ithaca, New York, was invited to Taiwan in 1950 jointly by the Economic Cooperation Administration Mission to China and JCRR. After reviewing the existing laws, regulations, and organization of the associations, he made a number of recommendations for their revision. To implement his recommendations, the Taiwan Provincial Government set up in January, 1951 an ad hoc committee composed of representatives of JCRR, the Provincial Department of Agriculture and Forestry, and the Provincial Farmers' Association, with the Provincial Governor as Chairman. Three months later, the committee submitted to the National Government a draft of a revised law governing the farmers' associations. But it was only after prolonged and thoroughgoing discussions by the authorities concerned, during which modifications were made in the original draft to reconcile conflicting interests, that the revised law and supplementary regula-

[3] Min-Hsioh Kwoh, *Farmers' Associations and Their Contributions Toward Agricultural and Rural Development in Taiwan* (Bangkok: FAO/UN Regional Office for Asia and the Far East, 1964), pp. 5–7.

tions were officially promulgated in August, 1953. Their main provisions are as follows:

1. Membership. The members of farmers' associations are divided into two categories, active and associate. A farm family which earns at least half its income from farming is entitled to have one of its men enrolled in the local farmers' association as an active member. Active members have the right to vote, to hold office in the association, and to use all its facilities. A farm family which earns less than one-half of its income from farming is entitled only to associate membership, which confers on its holder all rights enjoyed by active members except those of voting and being elected to office. Though associate members may be members of the board of supervisors, their number must not exceed one-third of the total number of supervisors.

The composition of farmers' associations reflects the various facets of agricultural life. Among their members are to be found landlords, owner-operators, tenant farmers, farm hands, and graduates of agricultural schools doing agricultural improvement work or teaching.

2. Management. The respective functions of the boards of directors and supervisors are policy-making for the one and auditing and general supervision for the other. All directors and supervisors serve gratis and cannot hold any other jobs in the associations. The manager appointed by the board of directors as the chief executive officer of an association is paid for his services, as they require his full-time attention.[4]

Soon after the promulgation of the revised law and regulations, the Provincial Department of Agriculture and Forestry in cooperation with JCRR and the Provincial Farmers' Association began active preparations for the reorganization of the FAs at all levels. Supervisory personnel and working staffs for the county and township farmers' associations were trained; posters, pamphlets, and radio broadcasts publicized the impending reorganizations; township committees were created to screen the qualifications of active and associate members; and, lastly, elections for members of the boards of directors and supervisors were held. The township committee was composed of the

---

[4] *Ibid.,* pp. 93–117.

township chief, the chairman of the township council, the incumbent chairmen of the boards of directors and supervisors of the township farmers' association, and a member of the township farm tenancy committee.

The elections were held from the bottom up: first in the small agricultural units, then in the farmers' associations at the township and the county levels, and finally at the provincial level. The entire work was successfully completed on February 18, 1954.

Since the reorganization, which was made possible by the thorough-going preparations and the understanding and support of the farmers themselves, the FAs have been under the control of bona-fide farmers. Unlike the situation under the Japanese administration of Taiwan, no government official can hold any position of responsibility, such as director or supervisor, in the associations. To supervise the extension services of the associations and give them advice and guidance, the Taiwan Provincial Department of Agriculture and Forestry has set up and maintained a Farmers' Organization Division, which serves as a liaison between the FAs and the government authorities.

As the members of farmers' associations at all levels elect their own chairmen, directors, and supervisors without outside interference, the FAs have become democratically controlled. At the same time their members can practice democracy by choosing the best men among themselves to direct and supervise the affairs of the associations and by taking an active part in discussions at various meetings which are held from time to time.

Services and Facilities

The farmers' associations in Taiwan perform a variety of services and provide different kinds of facilities, including rural credit and savings, extension services, sale and marketing of farm products, rural health and transportation services, promotion of rural industry, mediation and settlement of disputes between members, and sale of farm tools and implements, food, cloth and dresses, boots and rubbers, bicycle tires, radio sets, sewing machines, soft drinks, and canned fruits. At the same time, they also provide facilities for rice milling and for the storage of rice and fertilizers, and participate in crop and livestock improvement work, all for the government which entrusts them with such tasks. Altogether, they collected and processed 670,000 metric tons of rice and distributed 700,000 metric tons of fertilizers and 100,000 metric tons of feeds in 1967.

In order adequately to render these services, the farmers' associations have developed and maintained a number of facilities such as warehouses, processing machines, trucks, seed and animal breeding farms, fishponds, and jute-packing machines. In 1967 the assets of the FAs included 595 hectares of farm land, 103 livestock-breeding stations, 330 hectares of nursery fields, 1,254 rice warehouses, 704 fertilizer warehouses, 695 agricultural commodity warehouses, and 582 rice mills. Of the 341 township farmers' associations, 296 have a credit department each. These credit departments have a deposit of NT$3.79 billion on the one hand, and outstanding agricultural loans to farmer members of NT$3.84 billions on the other.

Since 1952 the farmers' associations have had the task of implementing the agricultural extension program. For this purpose they provide technical assistance to their members for the improvement of crop and livestock production. Under this program 88,955 adult farmers were enrolled in 4,869 farm discussion groups, 40,580 farm women took part in 2,304 home improvement clubs, and 67,420 farm boys and girls joined 4,901 4-H clubs in 1967. The farmers' associations spent about NT$96 million for extension services in that year.

All the township farmers' associations are conveniently located so that any one of their members can reach the association to which he belongs and return home in a few hours' time. In this way, a farmer can take a cartload of rice to his local association early in the morning, exchange it for fertilizers or other things there, and bring them back before the day is out. Organized by and for the farmers, the associations are ever on the lookout for the best means to serve their members. As grass-roots agencies, they are favorably situated to help carry out governmental policies, promote agricultural development, and work for rural improvement.

Though essentially similar to cooperatives in the services they render to their members, the farmers' associations in Taiwan differ in that they have both active and associate membership, whereas the cooperatives do not maintain such a distinction in membership.

The agricultural extension, marketing, and credit services rendered by the farmers' associations will be separately treated in Chapters 10, 17, and 19, respectively, of Part III.

Table 8-1 gives a general view of the growth of the farmers' associations from 1961 to 1967.

Table 8-1. Growth of the farmers' associations in Taiwan, FY1961–FY1967

	1961	1967
Number of FAs	340	364
Membership	786,129	878,651
Employees	8,098	10,913
Capital and reserves	NT$194,741,000	NT$694,222,348
Total marketing and supply business revenue	NT$683,295,000	NT$1,120,342,249
Value of total agricultural production	NT$24,427,000,000	NT$42,896,000,000
Extension expenditures from:	NT$81,782,000	NT$96,651,554
members' contribution	NT$9,987,000	NT$10,072,880
transfers from credit and marketing business	NT$12,733,000	NT$19,918,140
government subsidy	NT$22,042,000	NT$27,521,741
incomes from assets and production services	NT$38,536,000	NT$39,138,892
Deposits	NT$1,048,978,000	NT$3,792,717,435
Loans	NT$741,646,000	NT$3,843,405,039
Hogs insured	190,719	369,887
Profits earned	NT$20,244,000 (293 FAs)	NT$60,814,357 (337 FAs)
Deficits incurred	NT$8,947,000 (44 FAs)	NT$2,296,937 (16 FAs)

Source: Prepared by Farmers Service Division, JCRR, in 1968.

9. Agricultural Plans And Achievements

The part played by JCRR and other agencies in agricultural planning and the objectives for the development of agriculture in Taiwan have been explained in Chapter 4 under the section entitled "Joint Program Planning." In the present chapter, the successive Agricultural Four-Year Plans, which JCRR personnel have helped to formulate, will be discussed. But as the First and Second Plans carried out from 1953 to 1960 have been presented in some detail by the writer in a previous work,[1] only the third and fourth ones will be examined here and their results summarized.

However, a word about the Second Four-Year Plan is called for. In planning the agricultural sector of the 1957–1960 plan, Committee D of the Economic Stabilization Board set up a seventh working group of specialists for budget and economic analysis in addition to the six working groups that had been established at the time of the First Four-Year Plan to draw up development programs for food crops, special crops, forestry, fisheries, animal industry, and soil conservancy. The creation of the seventh group foreshadowed a new trend in economic planning in Taiwan. Whereas the emphasis was entirely on the achievement of certain production goals during the First Four-Year Plan, economic considerations began to occupy an important place in planning from the Second Four-Year Plan on. This new trend has been a distinguishing feature of all succeeding plans.

---

[1] T. H. Shen, *Agricultural Development on Taiwan since World War II* (Ithaca, N. Y.: Cornell University Press, 1964), pp. 62–77.

Third Agricultural Four-Year Plan, 1961–1964 [2]

In 1960 the commissioners and technical staff of JCRR participated in drafting and screening the various sections of the Third Agricultural Four-Year Plan, which was designed to promote the further development and better use of Taiwan's agricultural resources, including water resources, farm lands on the plains, marginal lands and foothills, forest resources, and fisheries.

The development of agricultural resources leading to the further boosting of crop, forestry, fishery, and livestock production was expected to serve a threefold purpose: (1) to achieve self-sufficiency in food supply and a better nutritional level for a growing population; (2) to coordinate with the United States aid program to gradually achieve economic self-support in Taiwan through an increase in both volume and variety of agricultural exports; (3) to advance the industrialization of Taiwan by supplying industry with requisite agricultural raw materials.

The estimated investment requirement for agricultural development during the plan period was about NT$8,026 million, plus an additional NT$300 million to be used as a farm credit revolving fund. Of this projected amount, 36.7 per cent was earmarked for the development of water resources, 36.7 per cent for the development of crops and livestock production, 14.2 per cent for the development of forest resources, and 12.4 per cent for the development of fisheries.

This investment was expected to increase agricultural production by 23.8 per cent by the end of 1964, with 1960 as the base year. This would involve an expected increase of 19.3 per cent in crop production, 19.3 per cent in forestry, 25.0 per cent in fisheries, and 36.3 per cent in livestock production in the next four years.

Fourth Agricultural Four-Year Plan, 1965–1968

As a result of the first three Four-Year Plans, the economy of Taiwan gradually evolved from a predominantly agricultural to a mixed agricultural and industrial one. Consequently, the Fourth Agricultural

[2] Agricultural Planning and Coordination Committee, Ministry of Economic Affairs, *The Agricultural Program under Taiwan's Third Four-Year Plan* (Taipei: Committee, 1961), pp. 1–22.

Four-Year Plan had to cover more subjects than its predecessors. In formulating the Plan, eight working groups were organized under the Agricultural Production Planning Committee of the Council for International Economic Cooperation and Development (CIECD), of which the writer has been a member. Another JCRR commissioner, Y. S. Tsiang, served as convener, and the head of the JCRR Office of Planning and Programming, W. M. Ho, served as executive secretary of the Committee.

Each working group was responsible for drawing up a plan for a specific aspect of agricultural development: food crops, special crops, forestry, livestock, fisheries, water and land resources, manpower, and economic analysis. The composition of the groups was similar to that of those under the previous Four-Year Plans. But this time, some foreign advisers were included among the 110 specialists.

In formulating their plans, the working groups were guided by long-range development projections to insure a degree of consistency among the various sectors, fields, and aspects of agriculture in the final Plan. Chart 9-1 gives a general idea of how agricultural planning was developed step by step.

To begin with, specific targets and production goals based on long-range development projections were worked out. The next step was to draw up sectoral plans product by product, regional plans, and supporting measures. In doing this, close coordination was maintained among agencies engaged in agricultural resources development, production and marketing of farm products, research, extension, financing and administration. At the same time, efforts were made to maximize coordination among the government agencies, agricultural enterprises, and farmers' organizations. In this way the fullest possible degree of integration was achieved.

Objectives

The objectives of the Fourth Agricultural Four-Year Plan were not only to continue to boost food production to meet the needs of the growing population and to improve the nutrition of the people, but also to supply raw materials for industries in support of industrial development, to promote export, to create employment opportunities for rural surplus labor, and to reduce the pressure of farm labor on

Chart 9-1. Outline of agricultural planning

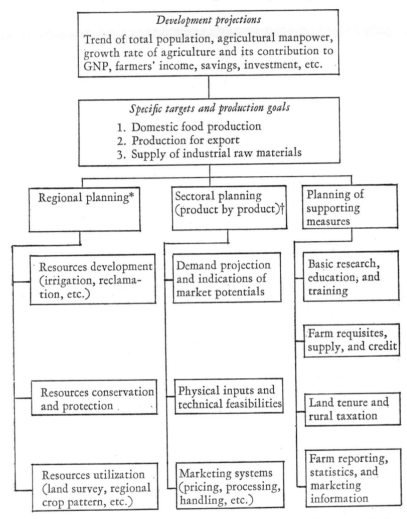

*Regions are classified by irrigation districts (or river basins), forestry districts, watershed area, tidal land, outlying islands, etc.

† It is necessary to formulate a sectorial plan for each agricultural product.

limited land resources. The measures taken to attain these goals were specified in the Plan.[3]

Crops

Since agricultural productivity in Taiwan was dependent on both land and labor productivity, steps were taken to develop land and water resources, adopt multiple cropping systems for intensive use of land, and improve the varieties of crops and farming techniques for increase of unit yield. Likewise, high priced products for export requiring intensive use of labor were developed as cash crops for farmers; year-round use of farm labor was encouraged to shorten the slack season; cooperative farming was promoted to remedy the defects of the small-farm system and to facilitate the introduction of mechanized farming.

Increase in the unit yield of rice was a chief concern of the agricultural workers, in as much as arable land on Taiwan is limited. Therefore, development of water resources and improvement of irrigation facilities were carried out to increase the acreage suitable for rice. Further development of miscellaneous crops called for the expansion of planted acreage through the adoption of multiple cropping systems. Varieties resistant to disease, frost, and drought were cultivated, using improved farming methods, to increase the unit yield of these crops. Farmers were encouraged to cultivate sweet potatoes in winter by the relay planting method in rice fields in the hope that the fallow acreage of rice in winter could be put to productive use. For peanut cultivation, emphasis was laid on developing river land and sandy land on the coast. Peanuts were grown by intercropping in canefields, tea plantations, and fruit orchards. Great efforts were made to expand the acreage of soybeans in winter and encourage farmers to grow them with tea, sugar cane, and corn by intercropping so that the legume crop might serve to improve soil fertility. Low-yielding areas of upland rice were shifted to growing corn. The corn acreage was further enlarged by the system of intercropping or by growing corn in paddy

[3] Council for International Economic Cooperation and Development, *The Republic of China's Fourth Four-Year Plan for Economic Development of the Province of Taiwan* (Tapei: Council, 1965), pp. 56–58.

fields in southern Taiwan with the relay planting method. Rape was planted in winter and grown on fallow paddy fields.

Among the export crops, sugar cane, tea, pineapples, bananas, citrus fruits, mushrooms, onions, asparagus, citronella, and fiber crops were grown on slopeland or low-grade land in plain areas. Crop rotation and intercropping systems were used to avoid competition with food crops for land use. The basic objective in promoting agricultural development in Taiwan was the achievement of maximum yield from individual crops and of highest returns from alternate crops.

Livestock

In livestock, the integrated swine improvement program was expanded to include food supply, breed improvement, and disease control. Draft cattle are still the major source of draft power on the farm, but their number is expected to decline after mechanized farming is introduced. Thus dairy and beef cattle are more important to the future economy. Farmers were encouraged to raise breeding heifers and calves on the rolling foothills in northern Taiwan, or to plant forage crops on upland and raise beef cattle. Demonstrations were planned, showing how pastoral agriculture incorporated in the various cropping systems can improve soil fertility. In poultry raising, efforts were made to improve breeds, feeding techniques, and disease control, and an integrated poultry production program was implemented to meet the demand for eggs and meat, as well as to promote the export of eggs, meat, and feathers.

To encourage livestock production, the feed industry and feed processing techniques were improved, along with the storage and processing of livestock products. Integration of sales and production was encouraged in order to stabilize prices for the benefit of both the producers and consumers.

Forestry

In the field of forestry, emphasis was placed on a well-balanced development of Taiwan's forestry program in which forestation, logging, utilization, and forest product processing and marketing were organized and coordinated. The foothills were cleared of understocked hardwood stands and replanted with desirable tree species. Attempts were made to raise economically valuable and fast-growing bamboo to meet the needs of the domestic and foreign markets. The huge volume of wood raw materials from forest conversion operations became available for pulp and paper, construction, or other kinds of manufacturing. The plan called for Taiwan's low-cost labor to be utilized in a well-organized and well-equipped woodworking and furniture industry for export. To meet the demand of these industries, forest product research was strengthened and expanded.

Efforts were directed toward reforesting areas with the species of trees needed by timber-using industries, as well as locating these reforestation areas in the neighborhood of these industries so as to lower the cost of logging and transportation. Timber-using industries were encouraged to operate their own forests and to promote contract reforestation.

In logging operations, priority was given to meeting the demand for industrial raw materials, to the improvement of the bidding system, and to the establishment of long-term supply contracts. It was planned that these measures would insure an adequate supply of industrial raw materials, so that forestry could support industrial development, and marketing of wood products could be expanded. Attention was given to the effect of logging on soil conservation; cutting was limited so that forests would continue to provide protection to the agricultural resources in the plain areas.

Fisheries

Sixteen tuna long-liners, constructed with loans from the World Bank, joined the overseas operations. Plans were made to encourage the establishment of a large deep-sea fishery corporation with the capability of competing in the world market, and to continue building tuna long-liners and large-sized trawlers for operating in offshore waters around Africa, Latin America, Australia, Samoa, and Penang Islands. Overseas bases for fishing operations were established. In inshore fishing, shrimp and other hitherto untapped resources were exploited; the use of such new gear as the fish finder increased the efficiency of the operation. In the matter of fish culture, measures were taken to increase the number of fish ponds and propagate high-priced fish products and shellfish for export.

Shore facilities were expanded to meet the needs of increasing production, and fuel consumption by the fishing vessels was cut to lower production cost. Attempts were made to improve market facilities, market management, and the marketing system in order to assure a stable price and higher income for the fishermen. Taking into account the growing demand in the world market for fish products, stress was put on surveys of the current trend in the world market, improvement of fish processing and freezing techniques, and strengthening of the foreign-trade organization with a view to increasing the export of fish products.

Water Resources Development

Ground water resources development was accomplished and plans were made for the development of land in east Taiwan, and in addition for that of the Wuchi-Choshui basin; and construction of the large-scale multipurpose Tsengwen reservoir was started. Irrigation and drainage systems were improved, with eleven projects scheduled for completion within the duration of this Plan. Ground water development was continued in central and southern Taiwan to expand the irrigation area and to increase food production. In land resources development and utilization, emphasis was placed on the development of tidal land along the western coast and on the use of river land in eastern Taiwan. To match the efforts in the development of slopeland, measures on soil conservation, proper treatment of illegally cultivated slopeland, settlement of reserved land for aborigines in mountain areas, and the planning and management of watershed received close attention. These measures helped to develop mountain agricultural resources and to bring about effective flood control. Projects for land consolidation, soil survey, and land improvement were continued.

The target for average annual growth rate of the agricultural sector during the four years was set at 4.1 per cent. Of this percentage, farm crops occupied a large portion, although the fact that available land was so limited adversely affected their development. Their average annual growth rate was 3.9 per cent. Since the yield of timber in 1964 (the year before the beginning of the Plan) was unusually high, attention to soil conservation was needed; for this reason the average growth rate for forestry was only 2.6 per cent. The average growth rates for livestock and fishery were 6.4 per cent and 3.3 per cent, respectively.

To attain these goals, the fixed capital investment required by the agricultural sector for the next four years reached NT$12,572 million at 1964 values. Of this amount NT$5,465 million were used for water and land resources development (43.5 per cent); NT$3,349 million for farm crops (26.6 per cent); NT$1,501 million for forestry (11.9 per cent); NT$1,521 million for fisheries (12.1 per cent); and NT$736 million for livestock production (5.9 per cent).

Achievements of the Plans

In spite of natural calamities such as typhoons, floods, and droughts, which are rather frequent occurrences in Taiwan and which human ingenuity has not yet been able to control, the successive Agricultural Four-Year Plans have been smoothly and successfully carried out and most of the production goals have been attained.

Based on the constant farm price of 1951, the average annual growth rate of agriculture, to which JCRR projects have contributed an important share, was 6.2 per cent from 1953 to 1956, 5.0 per cent from 1957 to 1960, 6.4 per cent from 1961 to 1964, and 6.6 per cent from 1965 to 1968 with an annual average of 6.0 per cent for the entire sixteen-year span.

The net result is that the aggregate agricultural output of crops, livestock, fisheries, and forest products in 1968 almost tripled that of the 1950–1952 average, or that of the prewar peak year as shown in Chart 9-2.

This significant increase in agricultural production has made it possible for Taiwan not only to provide the domestic food requirements of the population, which grew from eight million in 1952 to over 13 million in 1968, but also to leave a substantial surplus for export. The production of rice, the staple food of the people, increased from 1.6 million metric tons of brown rice in 1952 to 2.5 million metric tons in 1968. The increase in total rice production was due primarily to the increase in the per hectare yield from 1,998 kg. of brown rice in 1952 to 3,188 kg. in 1968, together with a slight increase during this interval in the area planted to rice from 786,000 to 790,000 hectares. The

Chart 9-2.

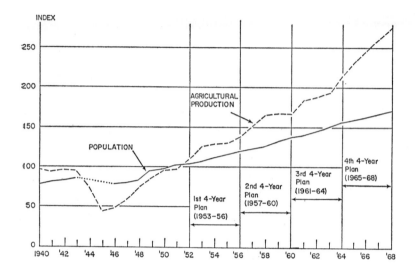

increase in the per hectare yield was, in turn, due largely to improved varieties, more and better use of fertilizers, irrigation, and pesticides, and better cultural methods.

It has been estimated that the per hectare yield of all crops increased by 26 per cent in the 1946–1952 period and by 59 per cent in the 1952–1967 period. The higher yields of crops other than rice was also due in the main to technological improvements and in part to soil conservation and land reclamation which had increased the total area of cultivated land from 876,000 hectares in 1952 to 902,000 hectares in 1967. With more food produced, the per capita daily food available steadily increased from 2,078 to 2,548 calories and the per capita daily protein intake from 49 to 68.2 gm. from 1952 to 1967.

The increase of agricultural production also boosted the net foreign exchange earnings from US$114 million in 1952 to US$292 million (excluding US$43.6 million of plywood and plywood products export) in 1967, through the export of sugar, rice, bananas, pineapples, tea, mushrooms, asparagus, fruits, vegetables, forestry products other than plywood and plywood products, and other primary processed agricultural products.

As agriculture and industry are basically partners, the development of the one in Taiwan has been linked with that of the other. Agricultural outputs such as sugar cane, mushrooms, pineapples, and asparagus become industrial inputs, while industrial outputs such as fertilizers, pesticides, and farm implements become agricultural inputs.

Production has been increasing in both agriculture and industry, with the latter forging ahead at a faster speed. From 1952 to 1968 the total agricultural production including crops, livestock, fisheries, and forestry tripled in physical terms. However, when expressed in percentages of the total net domestic product, the value of agricultural products declined from 35.7 per cent in 1952 to 23.2 per cent in 1968, while industrial products, including manufacturing, mining, construction, and electricity, increased from 17.9 to 29.2 per cent in this sixteen-year period.

Both agricultural and nonagricultural exports have been climbing. The value of agricultural exports tripled from 1954 to 1967. Yet in percentage of Taiwan's export, the agricultural exports declined from 93 per cent in 1954 to 44.9 per cent in 1967, while nonagricultural exports increased from 7.0 per cent to 55.1 per cent in the same years.

Fifth Agricultural Four-Year Plan, 1969–1972

The average annual growth rate for the Fifth Four-Year Agricultural Plan (1969–1972) is 4.45 per cent, which is slightly higher than the 4.10 per cent annual growth rate in the Fourth Four-Year Plan. This goal has been worked out after considering the present potentials of agricultural production, the demand of domestic and foreign markets, and future agricultural development plans.

Since cultivated land areas are limited, the annual growth rate of crops is planned at 3.13 per cent, of which rice is set at 2.32 per cent. Miscellaneous and feed crops could be further developed if the marketing facilities are improved and production costs lowered. The annual growth rate in forestry is projected at 3.63 per cent because of the low forest growing stock and the need of soil conservation in high areas. For fisheries and livestock, the annual growth rate is expected to reach as high as 12.6 per cent and 6.55 per cent respectively, since they are not restricted by land resources.

In the past years, the annual growth targets have usually been exceeded. However, this situation may not continue. Faced as Taiwan is by continued Communist threat, its military burden not only cannot be lightened, but may even become heavier. The rapid development of industry in the Fifth Four-Year Plan may occupy more of the cultivated land and absorb a larger labor force from rural communities. If this happens, agricultural growth will be seriously affected.

PART III. MAJOR PROJECTS

10. Agricultural Education, Research, and Extension

During the fifty years of Japanese occupation of Taiwan, elementary education was universal, but the local people were not actively encouraged or assisted to enter college, except for studying technical subjects such as medicine and engineering. Consequently, local leadership was rather undeveloped.

Only a very small number of the local people, who were all Chinese by blood relationship and cultural and historical heritage, had a chance to go to Japan or the Chinese mainland to receive higher education. This group hardly sufficed to fill the vacancies created by the repatriation of all the Japanese nationals at the end of World War II, when Taiwan was restored to Chinese sovereignty. This lack of administrative and technical personnel was made up by the large numbers of Chinese administrators, scientists, engineers, and agriculturists who came to Taiwan from the mainland either on government assignment or of their own free will. Together with the local leaders, these highly trained men undertook the task of rehabilitation and reconstruction in the first few years of the postwar period.

When JCRR headquarters was transferred to Taiwan in 1949, the Joint Commission immediately began to train the local people who were employed in the agricultural agencies which were asked to sponsor JCRR projects. Altogether, fifteen training classes, each of a specific nature and for a specific purpose, were conducted under the supervision of JCRR technical divisions in 1949 and 1950. The classes covered a wide range of subjects from animal husbandry and anti-malaria techniques to midwifery and veterinary practices.[1]

[1] JCRR, *General Report II* (Taipei: JCRR, 1951), p. 151.

To improve the ability and working efficiency of some 8,000 staff members of farmers' associations at all three levels, provincial, county, and township, JCRR also conducted a series of training classes[2] in preparation for the reorganization of the FAs, described in Chapter 8.

A Scholarship Program[3]

Besides conducting local training classes, in 1951 JCRR and the China Mission of ECA jointly sponsored a scholarship program, a long-range training project of broad scope. An eleven-member Scholarship Committee selected candidates for advanced training in the United States in the fields of agriculture, engineering, public health, natural sciences, and social sciences.

Most of applicants were recommended by colleges and universities and government agencies, and a smaller number filed their own applications. The examination was divided into three steps. First, a candidate's knowledge of English and Chinese was tested. Of the 709 candidates taking the initial examination, only 340 successfully passed the language test. Next, the successful candidates had to take a second written test on the technical subjects in which they specialized. This time only seventy-four of them emerged with satisfactory results. Finally, each one of this selected group was given an oral examination in the form of an individual interview on the subjects of the second written test and on his personal experience and knowledge.

Out of this series of tests thirty-six candidates were chosen, of whom ten came from mainland provinces and twenty-six from Taiwan. One of the successful Taiwan-born candidates was Henry Kao, the present mayor of the Special Municipality of Taipei.

After an eight-week orientation course, which included English, the thirty-six men were sent to the United States in the fall of 1951 to do postgraduate work in universities or to receive practical training in factories, hospitals, research institutes, or laboratories. All of them returned to Taiwan after their one-year training to continue to work for the organization or agency that had recommended them under the ECA-JCRR sponsored program, or to serve in an organization in

2 *Ibid.*, p. 41.
3 *Ibid.*, pp. 151–152, 217–223.

the field of their specialty for at least two years, as required by the program sponsors.

Strengthening Agricultural Education

JCRR has assisted in strengthening agricultural education in Taiwan. It has helped to improve basic training facilities in the Colleges of Agriculture of the National Taiwan University (NTU) and the Taiwan Provincial Chunghsing University.

On instructions from the Ministry of Education, the National Taiwan University appointed in June, 1953, a planning committee for the improvement and further development of its College of Agriculture, composed of two Chinese and one American commissioner of JCRR, the commissioner of the Taiwan Provincial Department of Agriculture and Forestry, and the dean of the NTU College of Agriculture, with the writer as committee chairman. The committee's recommendations dealt with improvements in curriculum, research, and training, and cooperation with some American institution of higher learning.

The committe first studied the existing curriculum of the NTU College of Agriculture, the composition of its faculty, the record of its students, the condition of its laboratories, the facilities of its library and experimental farm, and the college budget. Then it recommended that JCRR financial assistance be given to the college for the repair of its buildings, classrooms, offices, laboratories, and desks and chairs, the improvement of its water supply and its experimental farm, and the procurement of the necessary laboratory supplies, textbooks, reference books, and missing back numbers of technical journals.[4]

Following the planning committee's recommendations, JCRR invited Leland E. Call, former dean of the College of Agriculture of Kansas State College and Chief of the Agricultural Division, Mutual Security Agency (MSA) Mission to the Philippines, to make an on-the-spot study of the curriculum of the NTU College of Agriculture and to make suggestions for its improvement. Dr. Call came from Manila and stayed in Taiwan for two weeks in the summer of 1953 for this purpose. The recommendations he submitted to JCRR

[4] JCRR, *General Report V* (Taipei: JCRR, 1954), p. 94.

included the following principles: (1) courses should be offered on a term instead of on a yearly basis; (2) more general and basic courses should be required; (3) the number of credits to be accumulated through required courses should be reduced so as to enable the students to take more advanced courses offered as electives.[5] He also drew up a new curriculum for each of the nine departments of the College, as well as recommending the creation of several new divisions, including a "Soil and Fertilizers" Division and a "Food Technology" Division under the Department of Agricultural Chemistry, a "Plant Pathology" Division and an "Entomology" Division under the Department of Plant Pathology and Entomology, and an "Animal Husbandry" Division and a "Veterinary Medicine" Division under the Department of Animal Husbandry and Veterinary Medicine.

All of these recommendations were adopted with minor modifications by the NTU College of Agriculture.

Cooperation with American Agricultural Colleges

In response to another recommendation of the planning committee, Knowles A. Ryerson, dean of the College of Agriculture of the University of California, was invited to make a study of agricultural education in Taiwan and, if possible, to initiate cooperation between the two institutions of higher learning. Following Dr. Ryerson's visit, a three-year contract, signed in the fall of 1954, provided for the rendering of technical advice and assistance by the University of California College of Agriculture to its counterpart of the National Taiwan University in improving and strengthening its instruction and research.

Accordingly, a two-member advisory team came from the University of California in the winter of 1954 and immediately took up their duties at the NTU College of Agriculture. To insure the best possible cooperation between the American advisers and their Chinese colleagues of NTU, arrangements were made between JCRR and NTU for the loaning by the former of Paul C. Ma, head of its Plant Industry Division, to serve as dean of the latter's College of Agriculture. At this time a summer course was introduced for students who had completed their freshman year of studies to work and live on farms dur-

[5] JCRR, *General Report VI* (Taipei: JCRR, 1955), p. 134.

ing the summer vacation. In this way, the students gained some practical experience in farming before they continued their studies in the sophomore year.

With the adoption and implementation of its recommendations, the planning committee had served its purpose and was dissolved early in 1955.

The contract with the University of California expired in 1957. In the next three years, the NTU College of Agriculture carried on its teaching and research with continued JCRR technical and financial assistance but without the benefit of further aid from any American university.

However, as early as 1958, the idea of a cooperative program between the Michigan State University on the one hand and the NTU College of Agriculture and the Taiwan Provincial College of Agriculture (later incorporated into the Taiwain Provincial Chunghsing University as its College of Agriculture) on the other, was conceived. A group of Michigan State University professors made a study of the proposed program in 1959; an operational plan was drawn up, and a tripartite contract was signed by the Agency for International Development representing the United States government, the Ministry of Education and the Taiwan Provincial Department of Education representing the Chinese government, and the Michigan State University in July, 1960.[6]

The first MSU adviser arrived in Taiwan in October of the same year, and others soon followed. Up to the termination of the contract on July 31, 1964, eight two-year-term advisers and five short-term ones had been in Taiwan, seven of them working with the NTU College of Agriculture in Taipei and six of them with the Provincial Chunghsing University of Agriculture in Taichung.[7]

According to the terms of the contract, the MSU advisory team would assist in the development and strengthening of agricultural education at the two universities to provide adequately trained vocational agricultural teachers, supervisors, technicians, and farm leaders; fur-

[6] *Final Report of the Michigan State University Advisory Program to the Colleges of Agriculture of National Taiwan University and Provincial Chung Hsing University of the Republic of China and the Agency for International Development for the Years of 1960–64* (Taipei, 1964), p. 1.

[7] *Ibid.*, pp. 1–3.

ther develop and strengthen the basic agricultural instruction, research, and extension program of the two universities; and advise and assist them in their entire college program including administration, instruction, and procurement.[8]

It was also provided that the activities of the MSU advisory team would be carried out on an advisory and consultative basis, and through conducting seminars. In some cases the team members might undertake the teaching of certain classes. This would be done, however, only when a staff member of the two universities assumed joint responsibility for the organization and conduct of the course so that the local staff member could continue to teach that course in later years. In such cases, joint responsibility would include preparation of the course outline, lecture and laboratory materials, instructional aids, teaching, and examinations.[9]

Through both of the contracts with the University of California and the Michigan State University, several young professors and instructors were enabled to pursue advanced studies in the graduate schools of the two American institutions. After the completion of their academic work in the United States, all of them returned to their original posts in Taiwan, thereby greatly strengthening the agricultural teaching and research of the National Taiwan University and the Taiwan Provincial Chunghsing University.

As the American advisers conducted seminars, and taught only a few courses in exceptional cases, the undergraduates did not have the direct benefit of their teaching. Those directly benefited by their presence on the university campus were the Chinese professors and instructors. However, as a result of the services of the American advisers, there was a general improvement in agricultural education in Taiwan, as evidenced by the better quality of graduates turned out by the two colleges in recent years. A number of these graduates later went to the United States to pursue advanced studies and made an excellent showing in academic work.

To complement the work of the American advisory team, some JCRR senior specialists were permitted to teach undergraduate courses in the College of Agriculture of the National Taiwan University and

[8] *Ibid.,* p. 3.
[9] *Ibid.,* p. 5.

that of the Provincial Chunghsing University. At the same time, JCRR also extended financial assistance to the two institutions to improve their instruction and research. The amount appropriated by JCRR for agricultural research and education totaled NT$90,627,000 and US$731,200 up to the phasing out of United States economic aid on June 30, 1965, as shown in Table 5-5.

Assistance to Research

For the implementation of the two contracts mentioned above, JCRR assisted the National Taiwan University and the Provincial Chunghsing University in carrying out research, teaching, and extension projects all through the period from 1954 to 1965. In research, JCRR stressed projects that helped to promote agricultural development in Taiwan. The projects of the National Taiwan University receiving JCRR assistance in 1961 may be taken as an illustration of the kind of research that was encouraged.[10] In that year the NTU agricultural research program included the following topics: isotopic studies on plant nutrition, studies on marketing losses of vegetables, pathogenesis of edema disease of swines, studies on the Li-ku-bin disease of citrus, study of swine ascarid infestation through placenta, breeding and cytogenetic studies on sorghum, biochemical studies on mushrooms, studies on the interrelation of physiochemical properties and nutrient utilization of certain important zonal soils, studies on the mineral nutrition status of Taiwan citrus, and investigation of the preparation and packing of pickles.

In 1961 special funds were appropriated to strengthen the laboratory facilities of the National Taiwan University in the fields of soil and fertilizer, the anatomy of domestic animals, veterinary pathology, pesticide chemistry, biochemistry, and food technology. JCRR also financed the Audio-Visual Center and the Library of the NTU College of Agriculture.

As to the Provincial Chunghsing University, JCRR funds were made available in the same year for the building of the Animal Husbandry Hall, the Chemistry and Physics Hall, the Horticultural Hall, a greenhouse, and an insectary for its College of Agriculture.

[10] JCRR, *General Report XII* (Taipei: JCRR, 1961), pp. 163–164.

Similarly, JCRR helped to strengthen the teaching facilities and improve the quality of teachers in the provincial junior colleges of agriculture and the agricultural vocational schools.

Research by Other Agricultural Institutions

In addition to the National Taiwan University and the Provincial Chunghsing University, other agricultural institutions carry out agricultural research programs of their own, some of which receive JCRR assistance.

The Provincial Department of Agriculture and Forestry has under its jurisdiction four research institutes in charge of research programs in crops, livestock, forestry, and fisheries, respectively, and seven district agricultural improvement stations to carry out experiments and regional trials of crop varieties, cultural methods, livestock breeds, pest and disease control, fertilizer application, etc. The research institutes of the Department have field stations to carry out similar tasks.

In providing assistance to agricultural research by agricultural institutions, JCRR has chosen those projects that help to solve problems faced by farmers in general. Assistance to the research institutes and agricultural improvement stations of the Provincial Department of Agriculture and Forestry has been in the form of procurement of new equipment and supplies, financing of the construction of necessary buildings, provision of working funds for conducting experiments and surveys, and the advancement of experimental methods.

Research Personnel

There are approximately one thousand scientists and technicians engaged in research work in some sixty agricultural agencies. Of this number about 56 per cent are college graduates with academic degrees and 44 per cent, graduates of agricultural vocational schools. The fields they specialize in include crop production, plant protection, soil and fertilizers, farm machinery, animal industry, forestry, fisheries, food processing, and irrigation and drainage. The agencies for which they work are colleges and universities, specialized research institutes and experiment stations, agricultural improvement stations, and government and private laboratories.

Project Employees

In addition to its other contributions, the Joint Commission undertook to provide project employees for the research institutes, as shown in Table 10-1, whenever their budgets did not permit them to take on additional staff members.

The applied agricultural research work carried out by the local technicians with JCRR financial and technical assistance has contributed materially to the development of agriculture in Taiwan since the mid-fifties. The major practical results will be related in Chapters 11 to 20.

Table 10-1. Employees in JCRR-financed projects, March, 1966

Category	No. of employees	Educational level	
		College graduates	Senior vocational school graduates
Plant industry	406	150	256
Irrigation & engineering	35	17	18
Rural health	312	21	291
Agricultural economics	50	24	26
Forestry	218	119	99
Agricultural credit	5	5	0
Farmers service	16	13	3
Fisheries	11	1	10
Animal industry	34	15	19
Total	1,087	365	722

Agricultural Extension

During the Japanese occupation, extension work was carried out by government agencies at all levels. The farmers' associations then in existence also implemented some extension programs. But since the associations had offices only at the county level, the field work was channeled through local governments. In principle, the extension programs were mapped out to facilitate implementation of government policies rather than to directly promote farmers' economic welfare and, methodology-wise, they relied solely on subsidies rather than on educational means to develop farmers' decision-making capabilities. These early extension programs contrast sharply with the present ones, which are built on farmers' own organizations, aimed directly at increasing farmers' economic benefits, and employ educational methods to develop farmers' own initiative. In short, farmers are now the ends instead of means.

For the first decade of the postwar period, agricultural extension largely kept to the prewar patterns. Most of the extension personnel were part-time workers dealing largely with administrative routine; to improve their effectiveness, a new system of cooperative agricultural extension was introduced by JCRR into Taiwan during 1952 to 1956. Under the new system, the farmers' associations at all levels assumed the major responsibility of educating the farmers, with personnel and financial support from governments and with technical assistance from research and educational institutions. Relieved of administrative burdens, extension workers now became full-time teachers and consultants of the farmers' associations and thus were able to develop extension programs to meet the real needs of the farmers. The creation of the Agricultural Extension Office of the Provincial Department of Agriculture and Forestry in 1964 and the promulgation of the Provincial By-law Governing the Extension of Agricultural Extension Service in 1965 completed the present pattern of agricultural extension in Taiwan.

Agricultural extension in Taiwan includes three aspects, each serving one sector of the rural community. The first 4-H clubs for farm boys were organized by JCRR in 1952. Like their counterparts in the United States, the clubs taught the youngsters the rudiments of farming, under the guidance of local leaders and 4-H advisers. Similar clubs for farm girls were organized four years later. There are 4-H clubs in secondary schools as well as in villages. In 1968 fifty-eight schools and 304 townships had a total of 5,914 project clubs with 84,112 members.

If the 4-H clubs take in rural youths as members, adult farmers are given farm extension education through farm discussion groups, which numbered 5,351 with a total membership of 101,863 at the end of 1968. Initiated by JCRR in 1955, this aspect of agricultural extension has been carried out by farm advisers, who teach the farmers new farming practices, help them organize discussion groups, and hold demonstrations on selected farms. Audio-visual aids are used whenever and wherever possible.

The third aspect of agricultural extension in Taiwan, home economics training for farm women, was introduced in 1956. The work is implemented through home improvement clubs set up in various townships. In 1968, there were 2,636 such clubs with a membership of 46,522 scattered in 246 townships of the province.

When the agricultural extension service first began in the early fifties, each farm extension adviser was assigned to a definite area to work as a "generalist agent"; he had some degree of competence in his subject matter specialty, and he was buttressed by agricultural improvement station specialists. This type of service was suited to the less developed and less complex farming operations of the late forties and the fifties.

But the changing pattern of farming and the more complicated use of technology in pest control, fertilizer applications, irrigation water management, and other farming operations, has required a much better trained generalist extension worker with more effective help from subject matter specialists.

In recent years agriculture in Taiwan has become more diversified, with farmers raising several crops a year. Some newly introduced crops, such as mushrooms, asparagus, and onions, are being widely grown for export. Also more poultry, hogs, and dairy cows are being raised by farmers. In addition to a more diversified system of farming in use on many farms there is also emerging specialized farming in certain crops, such as fruits and vegetables, and in animal husbandry, particularly poultry and hogs. Highly competent agents are needed to advise these specializing farmers. As a result of general economic and social advances, the younger farmers in Taiwan, whether operating general diversified farms or more specialized farms, are better educated and more intelligent than their fathers and so require a higher quality of extension advisory assistance than in the past.

As agricultural technology continues to improve and farming operations become commercialized, the fact that farms cannot be made larger poses more of a challenge than ever before. JCRR Commissioner Gerald H. Huffman pointed out in a note to the author: "Taiwan has a more difficult problem of deciding on the future township farm extension adviser organizational pattern than the United States. In the United States our farms are becoming larger and they are moving away from diversification to specialization. Consequently, the local extension organization is changing from generalist county agents to specialized subject matter specialists at the area or multi-county level. These specialist area agents in a specific field work only with the farmers in their area who concentrate on one or more farm enterprises."

In Taiwan, however, there are two distinct trends in farming occurring at the same time—diversification and specialization. The farmers working on the specialized farms are really specialists themselves, and a generalist local farm agent just cannot keep ahead of them. For the diversified farm the problem is more difficult. In this case the question is whether attempts should be made to upgrade the training of the generalist agent and provide him with better subject matter backstop, or whether the local staff should be more and more specialized, requiring several of these specialist local agents to work with the same farmer. Generally speaking and from a farm management point of view, it would seem better not to have several local farm agents advising the same farmer. Moreover, this procedure would be too costly in view of the large number of farmers to be served in each township.

In the light of the foregoing, it would be advisable for JCRR to follow a twofold approach in agricultural extension. On the one hand, the training of generalist local agents should be upgraded and more effective subject matter specialist support should be provided for them, and on the other, specialized local agents at the multi-township or county level should be trained to work directly with the more specialized farmers and also to provide part of the subject matter support for the township generalist agents working with the diversified farmers.

The challenge of agricultural extension work is to keep abreast of the rapidly changing needs of individual farmers and the production and marketing problems each farmer must solve. This dynamic situation requires that periodic reviews and adjustments be made in the staffing pattern of local agents who are in direct contact with the farmers. Otherwise, the extension service will atrophy and cannot perform the functions for which it has been established. JCRR's goal in farm extension is to keep this program attuned to the present rather than to the past.

11. Land Use and Soil Conservation

The small area of cultivated land, about 883,000 hectares in all, has been a limiting factor in crop production in Taiwan. Arable lands on the plains are intensively cultivated, and the high mountains are covered with forests. It is the zone in between that has presented many land-use problems. Vast areas of low hills are covered with various hardwoods of low economic value, or used for the growth of tea, bananas, pineapples, sweet potatoes and other crops; but soil conservation has been neglected.

To promote better land use, JCRR initiated a marginal land survey in 1952, a soil conservation program in 1953, and, for the plain areas, a land consolidation program in 1959.

As the land use and soil conservation program in Taiwan has been treated in a previous work by the writer,[1] only the later projects of JCRR will be presented here together with a brief hint of the work done earlier.

Marginal Land Development

Better use of the crop-forest marginal land has been a long-range goal of agricultural resources development on the island. For the attainment of this goal JCRR has helped carry out a number of projects including: (1) a comprehensive land-use survey of slope lands; (2) training of personnel and demonstration of soil conservation principles and methods; (3) introduction of and experimentation with deciduous fruit trees; (4) development of grazing lands, and (5) ac-

[1] T. H. Shen, *Agricultural Development on Taiwan since World War II* (Ithaca, N.Y.: Cornell University Press, 1964), pp. 104–108.

celeration of reforestation. The last three projects will be discussed, respectively, in Chapters 13, 14, and 15 of this work.

The soil conservation survey begun in 1953 was the first land-use project ever carried out in Taiwan. It covered all the crop-forest marginal lands lying below 1,000 meters in altitude in Taiwan and its purpose was to collect data on which to base plans for the future use of these lands.

The survey was first suggested by Tom Gill,[2] who served as a short-term forestry consultant to JCRR in 1952, and developed by Raymond H. Davis, JCRR commissioner (1952–1959). Using aerial mosaics as base maps, the field work was started in 1953 by the Provincial Department of Agriculture and Forestry and completed in 1958; land-use capability maps for about 1,500,000 hectares were published two years later. C. L. Orrben, then a soils and fertilizers adviser with the MSA Philippine mission, came to Taiwan several times at the invitation of JCRR for consultation on the technical aspects of the program. The lands were divided into eight classes, totaling 1,313,754 hectares.

To this figure there should be added 184,131 hectares of paddy fields, which present no soil erosion problems, so that the total area surveyed would be some 1,500,000 hectares.

For rationalizing land use in the marginal zone, JCRR, in cooperation with the Provincial Department of Agriculture and Forestry, has been endeavoring to improve soil and crop management of the slope lands, give protection to and reclaim the idle or otherwise not properly utilized lands, and promote the conversion of scattered tree areas to forests of high economic value.

Farmers in Taiwan formerly planted windbreaks, built revetments, and followed some other simple practices to protect their lands from the ravages of nature. However, it was not until 1953 that efforts were made to deal with the problem of soil conservation in a systematic way. The early work, which was promoted by JCRR, in cooperation with the Provincial Department of Agriculture and Forestry, consisted mainly of demonstrations on improved conservation methods and training activities carried out by local soil conservation field offices. These efforts laid a good foundation for the later development of

[2] Tom Gill, *A Forest Policy and Program for Taiwan,* Forestry Series No. 2 (Taipei: JCRR, 1952), pp. 8–12.

the soil conservation program which began in 1961 with the establishment, under the Provincial Department of Agriculture and Forestry, of the Mountain Agricultural Resources Development Bureau. In addition to soil conservation, the bureau is responsible for the development of grasslands and other agricultural resources in the mountain areas.

Large-scale Soil Conservation

In 1954–1961 I. K. Landon, JCRR consultant, trained a large number of soil conservationists and developed competent leaders, and then carried out a four-year large-scale soil conservation program in 1963 to 1966. Though the original goal was to initiate conservation practices on only 45,000 hectares, finally 46,125 hectares of hilly farm lands, including private lands, public lands, and aboriginal reservations were included.[3] The primary treatments were bench terracing, hillside ditching, stone walling, grass barrier setting, and land reclamation. However, other conservation practices such as green manuring, cover cropping, composting, mulching, revegetation, waterway and terrace riser stabilization were also applied, wherever and whenever needed.

Bench terracing is generally used on slope farm land. But owing to its high construction cost and the shortage of labor in rural areas, it is now being replaced by some less expensive but equally effective measures such as grass barriers, cover crops, and mulching on steep slopes.

The Integrated Soil Conservation and Land Use Program

To assist the Provincial Government to promote the development of slope lands and to strengthen the effectiveness of soil conservation, JCRR launched an integrated soil conservation and land use program in 1966.[4] Three places in Hsinchu, Hualien, and Taitung with a total of 1,090 hectares of slope land were chosen as sites for this project, with JCRR financing the operational expenses. Designed primarily for experimental and demonstration purposes, this program is expected to yield the following benefits:

1. Surveying, planning, and supervision are greatly facilitated at this

[3] JCRR, *General Report XVII* (Taipei: JCRR, 1966), p. 14.
[4] *Ibid.*, pp. 14–15.

stage by having the treated areas concentrated in one place. In the future it will be possible for field technicians to carry out their duties of soil conservation with greater ease and efficiency.

2. The coordinated planning of farm roads, farm ponds, irrigation and drainage systems, windbreaks, soil-saving dams, and other public facilities on a community-wide basis will make soil conservation treatments more effective, and the farmers will be able to cultivate their lands more intensively.

3. Lands will be used according to their capabilities for raising crops recommended by experts. In this way the farmers will have higher and more dependable incomes.

4. The total expenses incurred by farmers on farms in the treated areas can be compared with those incurred by operators on nearby farms without the benefit of soil conservation; the comparison will point up the difference in the incomes derived from both types of farms. From this a pilot pattern can be worked out for future guidance.

5. The areas that have had the benefit of soil conservation will supply models for purposes of demonstration, education, and publicity.

The integrated soil conservation and land use program is one of the most important for agricultural development in Taiwan. It is a continuing program for improvement. As soil erosion is also serious in many developing countries, especially those in tropical and subtropical areas, Taiwan's experience in this respect may be useful to them.

Watershed Management

Watershed management deals primarily with the protection and conservation of water resources in the upstream areas. Activities in an upstream watershed such as cultivation, highway construction, logging, and mining may affect the quantity and quality of water. In Taiwan where the problems of sedimentation and floods are serious, recent efforts to develop mountainous lands to increase agricultural production have caused considerable damage to the downstream areas and reservoirs below. Unless man-made erosion is controlled and natural erosion is minimized, the development of water resources will be uneconomical in most parts of Taiwan.

In addition to the large-scale soil conservation and the integrated land use programs, much attention has been given to the protection of the major and reservoir watersheds in recent years. Watershed surveys were carried out at Shihmen (Taoyuan), Tsengwen Chi (Chiayi and Tainan), Paiho (Tainan), Houlung (Miaoli), Hualien Chi (Hualien), Mukua Chi (Hualien), Wusheh (Nantou), Touchien Chi (Hsinchu), Hsin-tien Chi (Taipei), and Pakuashan (Changhua) with JCRR's technical and financial assistance. In 1967 a preliminary survey of Taiwan's watersheds by aerial photogrammetry was completed. The findings provide data for overall watershed classification and management. Other surveys on specific watershed problems such as landslide situations in the basins of Choshui Chi and Wu Chi (Central Taiwan), the condition of existing check dams in Taiwan, and the distribution of the severely eroded mudstone regions have also been completed.

Field watershed management activities such as slope stabilization, gully control, check dam construction, revegetation and conservation of cultivated slopes are carried out in most of the watersheds following the completion of the survey. Very recently, special attention has been given to the control of mining waste in the municipal watershed of Taipei, application of new techniques for stabilizing highway banks in central Taiwan, pilot treatment of mudstone areas in the southwest, and experimentation with new types of watershed structures and farm ponds.

A watershed management seminar with 39 participants from 16 government agencies and universities was held in 1967. It was the first seminar of its kind in Taiwan. Hydrologic studies of small watersheds were conducted at Lukuei (Kaohsiung) and Lienhuachi (Nantou), both on lands owned by the Taiwan Forestry Research Institute. Altogether nine watersheds were gauged and data collected. After a few years of calibration, the existing trees will be cut and the land will be shifted to other uses. The changes in the hydrologic and sediment behavior, if found statistically significant, will provide a basis for the overall watershed management program.

JCRR has also stimulated and participated in the preparation of a set of watershed management regulations for the Shihmen watershed. These regulations may be promulgated in the near future. Since 1963, JCRR has assisted actively the Forestry, Water Conservancy and

Soil Conservation Joint Technical Committee of the Taiwan Provincial Government in promoting coordination and carrying out joint watershed management projects.

Land Consolidation

Land consolidation may be considered a follow-up of the land reform program which has been successfully carried out in Taiwan. Before 1955 JCRR's land reform activities were primarily aimed at improving the farm tenure system and creating owner-operators. After that, emphasis was laid on the promotion of better use of agricultural lands. Hence, it was considered necessary to improve both the farm structure and the land structure by consolidating the fragmented farm holdings into plots of bigger size and more regular shape and making them directly irrigable, drainable, and accessible. This would lead to increased unit area yields and reduced production cost, and bring greater income to the farmers.

A pilot land consolidation project was carried out in 1959 by the Tainan County Government with JCRR assistance. It was implemented on 188 hectares of farm lands in Tachia, Tainan County, in close coordination with a plan for irrigation improvement. The first step was the construction of irrigation works. Then came the building of farm roads designed to fit into the consolidation pattern. Finally, old farm plots were lumped together and consolidated into new rectangular plots to be re-allocated to the landowners according to the size of their original holdings.

The Tachia project demonstrated that a large-scale land consolidation program could be worked out and implemented in Taiwan without technical difficulty and with the full cooperation of the farmers. Between 1959 and 1961, twenty-two projects covering 4,600 hectares of farm lands were carried out for demonstration purposes, with JCRR playing an active role in planning and implementation. Encouraged by the success of these projects, the Taiwan Provincial Government launched in 1961 a land consolidation program of its own, which was aimed at consolidating 300,000 hectares of farm lands in ten years. Thus, land consolidation in Taiwan finally moved from the demonstration to the extension stage. Up to the end of 1968, approximately 161,269 hectares or about 53 per cent of the 300,000-hectare goal was consolidated.

12. Water Control and Utilization

Taiwan is subject to typhoons and strong tropical storms almost every year. Heavy rainfall generates floods that are disastrous because of the unfavorable geological and topographical conditions. Floods wash out dikes, inundate lands, take lives, and destroy property. Flood control is naturally a matter of prime importance.

Flood control requires the thorough regulation of rivers and careful maintenance of dikes to safeguard agricultural lands and protect people's lives and properties. As for irrigation development, it calls for the continual development and better utilization of both land and water resources to boost agricultural production and promote general economic development.

In 1942 irrigated lands in Taiwan totaled over 500,000 hectares, or approximately 61 per cent of the cultivated areas on the island; there were 3,493 irrigation canals managed by 39 local irrigation associations, and the annual production of brown rice was 1,400,000 metric tons. But due to the neglect of maintenance and destruction of irrigation facilities during World War II, the area under irrigation was reduced to 298,765 hectares, and the annual brown rice production dropped to 638,828 metric tons in 1945.

Up to 1945, a total of 395 kilometers of levees had been built on the nineteen principal rivers, whereas only 23 kilometers had been constructed on the thirty-two secondary rivers in Taiwan. The benefited areas totaled about 200,000 hectares, including about 25,000 hectares reclaimed from river beds. However, these flood control facilities were damaged during the war, and money to repair them was not available. After the restoration of Taiwan to Chinese sovereignty it was found that more than 47.7 kilometers of levees had been destroyed and over 260,000 hectares of irrigated lands had been exposed to flood menace.

Policy Modification

Flood Control

The biggest flood in recent years occurred on August 7, 1959. The downpour lasted for days and was so intense that the flood water exceeded the maximum capacity of the rivers and artificial drainage channels by 60 to 130 per cent. The affected areas extended from Miaoli in the north to Tainan City in the south. At many places good lands were buried under layers of sand and gravel brought down by the floods, while elsewhere many dikes were washed away and the floods rushed in to inundate the fields. The floods also caused drainage problems, and large sheets of water remained on the land for many weeks. Great damage was done to Taiwan's economy.

Rehabilitation measures were taken immediately and, at the same time, careful investigations to find out why such an unprecedented flood could have happened uncovered two facts. First, damage was much more serious on unregulated and on poorly maintained rivers. Second, on rivers that had been divided into principal and secondary reaches for regulation, the damage was more serious on the secondary than on the principal reaches. These facts indicated the need for revising the old flood control policy. The new policy which took shape in later years includes several points:

1. *Overall Planning for Flood Control Works.* Each river is considered as a unit in the planning. The former practice of dividing a river into principal and secondary reaches is abandoned.

2. *Basin-Wide Regulation of Rivers.* In case of financial difficulty, the regulation work may be carried out by stages. But the work must be done by the joint efforts of the provincial and local authorities. The former practice of partial regulation is abandoned.

3. *Management and Maintenance.* A set of *Regulations Governing the Management of Rivers in Taiwan* was promulgated in December, 1965 by the Taiwan Provincial Government on the initiative of JCRR and the Provincial Water Conservancy Bureau. The Regula-

1. Farm lands in Taiwan were originally small fragments of irregular and scattered plots without adequate irrigation and farm roads. A consolidation program has remodeled these plots into larger areas, providing roads and means of direct irrigation and drainage.

2. Canal lining, a part of water resources development, has saved millions of cubic meters of water for irrigation of rice and other crops.

3. The Shihmen Reservoir, initiated by JCRR, is a multi-purpose water resources development with the four major functions of irrigation, power generation, flood control and public water supply.

4. A new, high-yielding rice variety. Such varieties, developed in Taiwan, are being used in many parts of the world.

5. Watermelon, a special crop grown all year round in Taiwan. The island has become the only country producing seedless watermelon for the international market.

6. Aerial application of pesticides is used for the control of highly transmissible and destructive pests.

7. A farm machine in use. The rapid mechanization of farms is freeing labor for industry.

8. Cattle raising for beef, a new industry. There were no beef cattle breeds on Taiwan until JCRR in 1962 imported ten Santa Gertrudis bulls to improve the native yellow cattle.

9. Hog production in Taiwan is next to rice production in importance. Improved breeds with hybrid vigor mean faster growth and better quality meat.

10. Small craft in Taiwan's off-shore fishing fleet. The JCRR fisheries program includes technical and financial assistance for production, processing and marketing of fish, as well as research and extension work.

11. A mountain railway. Mountains constitute two-thirds of the total area of Taiwan. To make accessible the abundant forest resources in these mountains, many forest roads and railways have been built.

12. A field day, promoting farm team work. These days have been conducted throughout Taiwan.

13. A 4-H Club farm project. Close to 85,000 4-H club members learn the rudiments of farming under the guidance of local leaders and 4-H advisers.

14. A farmer who obtains credit and supplies from his Farmers' Association. Total capital of these associations exceeds NT$694,000,000.

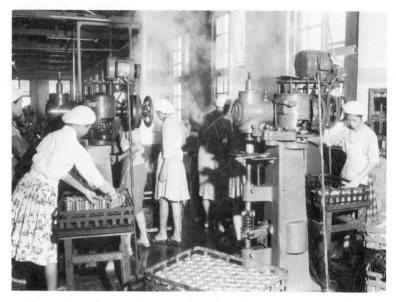

15. Processing of agricultural products in Taiwan is now a major industry.

16. A road on Quemoy. This once barren island has become green and productive. Well-patterned farm windbreaks and seacoast forests have been planted and the roadsides are lined with trees.

tions, which cover all aspects of flood control such as planning, regulation, utilization, maintenance, flood prevention, flood fighting and financing procedures, strictly adhere to the principles in points one and two.

Irrigation

Equal emphasis has been given the development of potential water resources and conservation and better utilization of available water.

Development work covers basin-wide water resources planning, construction of reservoirs, ground water development, and extension and improvement of existing irrigation facilities.

To achieve more effective use of water, projects have been implemented for canal lining to cut down percolation losses, and rotational irrigation has been promoted to expand the irrigated area.

Irrigation systems in Taiwan are managed by the irrigation associations. Good irrigation management depends much on the proper operation of the associations themselves. The organization of the associations has been strengthened from time to time. The most recent improvement was made possible by the promulgation of the *General Regulations Governing the Organization of Irrigation Associations* by the Central Government in July, 1965.

The Joint Irrigation Fund

The flood of August 7, 1959, damaged many irrigation systems. The rehabilitation work was entirely financed with JCRR loans and Government grants. The irrigation associations had no reserve funds in their budgets to meet such an emergency. In view of this fact, the Provincial Water Conservancy Bureau initiated and established in 1961 the Joint Irrigation Fund, with the director, chief engineer and head of the Water Administration Section of the Provincial Water Conservancy Bureau and chairmen of the 26 irrigation associations serving on its supervisory committee. A target of NT$500 million was set for the Fund. Deposits to the Fund by the associations were made possible through an increase in their membership fee in 1960 and through the reservation of a certain percentage of their membership fee collections in subsequent years. At the end of 1967, the accumulated deposits reached NT$421.9 million.

A nine-man standing committee was elected from among the members of the Supervisory Committee to administer the Fund and review loan requests. Loans are made from the Fund, at 6 per cent per annum, for emergency repairs and damage rehabilitation. New construction projects may be financed if there are sufficient available funds. At the end of 1967, loans totaling NT$447.6 million were extended for irrigation projects, and outstanding loans amounted to NT$253.4 million.

During 1961–1967, the operation of the Fund was so successful that it was no longer necessary for the Government to subsidize rehabilitation projects, and, as a result, more government funds were available for new construction. The Joint Irrigation Fund has provided a reliable source of financing for Taiwan's irrigation development.

The Flood Control Fund

The dikes in Taiwan have a combined length of about 700 kilometers, and most of them are badly in need of proper maintenance. More dikes, with an estimated combined length of 500 kilometers, are planned. However, the yearly appropriation of the Provincial Government for flood control is far from enough for both maintenance and systematic construction.

At the suggestion of JCRR, with details worked out by the Provincial Water Conservancy Bureau, the Provincial Government announced in 1965 a plan for the establishment of a NT$1-billion Flood Control Fund. Contributions to the Fund have so far reached NT$800 million, comprising NT$500 million from the Provincial Government in the form of stock shares of a public enterprise, NT$200 million from JCRR, and NT$100 million from the Central Government.

The Fund is composed of two parts, a Revolving Fund and an Endowment Fund. Loans from the Revolving Fund, at the interest rate of 6 per cent per annum, are for the construction of new regulation works. Such loans are to be repaid with the assessments collected from the beneficiaries. The Endowment Fund forms part of the stock shares of the Taiwan Fertilizer Corporation and therefore earns dividends, which are used for financing the annual maintenance of existing flood control works.

The Flood Control Fund has been in operation for three years under the supervision of a committee of the Provincial Government, of which JCRR is a member. To date, only limited amounts of funds have been released from the Revolving Fund, but as soon as the newly established River Regulation Planning Team of the Provincial Water Conservancy Bureau has drawn up feasible plans, larger sums will be needed from the Revolving Fund. On the other hand, the sizable dividend earnings from the Endowment Fund have greatly facilitated the maintenance of existing river works.

The Role of JCRR in Water Resources Development in Taiwan

Owing to the unfavorable meteorological, geographical, and hydrological conditions, and the neglect of proper maintenance and repair by the Japanese Colonial Administration during World War II, many irrigation and flood control facilities in Taiwan needed to be rehabilitated after the war. JCRR, after it took a hand in Taiwan's irrigation development, stressed chiefly the rehabilitation and improvement of the existing systems. It was deemed wiser to bring more lands under irrigation through the economical use of water available in canals and through avoiding its misuse than to develop new sources at great expenses. During the 1949–1952 period, 260,000 hectares of irrigated land was recovered for rice production with JCRR assistance.

In fiscal year 1952, the Council for United States Aid requested JCRR to provide technical supervision over the Provincial Water Conservancy Bureau's flood control projects that were financed by counterpart funds. However, the flood control program in this year and the following several years consisted mostly of dike rehabilitation construction or repair projects that only served to protect localized areas. It was realized that, without an island-wide program, water resources in Taiwan could not be fully developed and flood disasters could not be prevented. Consequently, data was collected for developing a comprehensive plan into which the local irrigation and flood control projects could fit. Therefore, while JCRR was taking up irrigation, drainage, and flood control projects to meet the immediate needs of farmers, it was mindful of the possibility of incorporating these projects in a multi-purpose basin-wide water development program or an integrated river regulation plan on a basin-wide basis.

The need for integrating new projects with old ones and with other possible multi-purpose projects resulted in the implementation of a project in 1952 for making reconnaissance surveys of the water resources of the various river basins in Taiwan under the joint sponsorship of the Provincial Water Conservancy Bureau and the Taiwan Power Company. These surveys provided most of the geological and hydrological data required for development planning for every watershed, and information on the status of existing projects on irrigation, power development, and flood control. They also indicated potential dam sites for future development.

Since then, a number of investigation and research projects have been carried out with JCRR assistance by the Provincial Water Conservancy Bureau, the irrigation associations, the Taiwan Power Company and the Water Resources Planning Commission of the Ministry of Economic Affairs. In 1952, US$12,621 was made available to support for the first time a research project for investigating, surveying, and model-testing the Peikang River, which eventually led to the formulation of the Peikang River regulation and flood control program and its implementation by the Provincial Government. JCRR assistance was also rendered for the Yenshui Chi Diking Project. The Yenshui Chi, a secondary river in the densely populated Tainan area, where intensive farming was practised by the farmers, often overflowed and caused heavy damage to the adjacent 2,800 hectares of land, and to railroads, highways and irrigation systems. The Provincial Water Conservancy Bureau was helped to formulate a plan of regulation and set up a three-year program (1960–1962) with a budget of NT$54,000,000. One half of the amount, or NT$27,000,000, was provided in the form of a grant. These examples illustrate the important role JCRR played, and the principles it adopted in water resources development in Taiwan.

On the mainland, it had been the policy of the Joint Commission to give financial assistance in the form of loans for irrigation engineering projects with the understanding that the irrigation districts concerned would repay these loans in annual installments at an interest rate of 6 per cent per annum. The repaid loans would go into a revolving fund for aiding similar rural engineering projects. Hence, each engineering project to be given aid must be self-liquidating and sound from both the technical and economic points of view.

The same policy has been adopted in Taiwan for irrigation, drainage, and reclamation projects except in the early 1950's; at that time rehabilitation and repair were so badly needed to restore agricultural production that assistance was given in the form of grants-in-aid in certain cases so as to reduce the financial burden of the farmers. By the time the NT$60-million Tapu Reservoir was built, financial aid was given on a 50 per cent loan and 50 per cent grant basis.

To encourage the farmers to begin using rotational irrigation, JCRR at first paid half of the cost of a project with a grant, and the other half was met with a loan to the local irrigation association; the percentage of grant has been gradually reduced to the present 30 per cent. Except for rotational irrigation projects, practically all JCRR financial aid for productive engineering projects is now given in the form of loans. Meanwhile, the interest rate on irrigation loans was raised to 12 per cent per annum from 1960 to 1965 and later revised to 10.08 per cent, 8.28 per cent, and 6 per cent per annum according to the nature of projects, as stipulated by the regulations of the Sino-American Fund for Economic and Social Development.

JCRR provided grants for all flood control projects through the Government Budget Supporting Program. However, since the establishment of the Flood Control Fund, no more such grants have been given.

Reorganization of Provincial Water Conservancy Bureau

The Provincial Water Conservancy Bureau is the agency responsible for flood control, irrigation, and drainage under the Taiwan Provincial Government. Established on July 1, 1947, it was originally composed of three technical sections, a management section, a procurement and supply section, and an administration section. The three technical sections were those on flood control, irrigation, and survey and drafting. The division of labor was so clear-cut that each section had its own group of engineers for planning, designing, construction, and awarding of engineering jobs of more or less the same nature. This was not an efficient way of using engineering manpower. Furthermore, the over-emphasized distinction between flood control and irrigation was contrary to the modern idea of developing a river basin as a unit.

After studying this problem for years, Thomas R. Smith, then chief of the Irrigation and Engineering Division of JCRR, prepared a plan in 1955 for effecting institutional improvement of the Provincial Water Conservancy Bureau. The three technical sections under the plan would be reorganized into one on planning, one on design, and one on construction. They would have the same overall objective —to promote water resources development covering flood control, irrigation, drainage, and storage—and their work would be interlocked and made interdependent. In addition, the plan also suggested that the control of water rights, formerly in the hands of the management section, be transferred to an independent Water Rights Office under the direct supervision of the Director of the Water Conservancy Bureau and in coordination with the technical and management sections concerned.

The JCRR plan was accepted by the Taiwan Provincial Government, and in May, 1956, the Provincial Water Conservancy Bureau was reorganized accordingly. The Provincial Water Conservancy Bureau has shown improvement since then. It has become more efficient and has been able to assume greater responsibilities in executing water resources programs on a basin-wide basis.

Improving Management of Irrigation Associations

During the past 20 years, JCRR assisted in the reorganization of the original 39 irrigation associations into 26 associations to improve their management. Following the reorganization in 1956, efforts have been made to improve the financial operation and irrigation service, and strengthen the activities of the irrigation groups of the irrigation associations.

The 26 irrigation associations in Taiwan collect regularly from the farmer members water fees to finance the maintenance and construction of irrigation facilities. In the past, most irrigation associations had no sound collecting procedures. This created loopholes and irregularities. To improve this situation, the Provincial Water Conservancy Bureau worked out in 1957 a new procedure for collection under the supervision of a designated agency—the Land Bank of Taiwan. The work of introducing the new procedure to replace all old practices was carried out in 1958 under the supervision of the Provincial Water Conservancy Bureau with JCRR participation and assistance. As a result, an improved system of collecting membership dues and auditing the accounts was established for all the 26 irrigation associations.

Most irrigation associations did not have accurate records of the irrigated lands on the basis of which water fees could be collected. To remedy the situation, a province-wide survey of about 500,000 hectares of irrigated land serviced by the 26 irrigation associations was conducted from 1963 to 1965 by the irrigation associations under the supervision and technical and financial assistance of the Provincial Water Conservancy Bureau and JCRR. The survey produced five million land cards and 10,000 irrigation maps which are now used not only for collecting water fees but also for planning irrigation and water resources development. In 1969 a follow-up project was scheduled to survey lands irrigated by private canals. This survey has provided comprehensive irrigation data that can be used to further improve irrigation management and plan future development programs in Taiwan.

The irrigation associations were also blamed for being overstaffed and for their inadequate service. To correct such defects, attempts have been made during recent years to strengthen the irrigation associations' services to farmers by reorganizing the 4,038 irrigation groups and 21,690 teams working in the 26 irrigation associations and strengthening their activities. Each irrigation group consists of farmers within an area of 50 to 150 hectares. The main function of the group is to help the irrigation working stations provide better irrigation service to farmers. Since the irrigation groups and teams were not an organic part of the irrigation associations and their members did not receive any pay, their performance did not come up to expectations.

Starting in 1965, a reorganization and strengthening program was implemented. The operating area of each group was readjusted to a more workable size. Young farmers were elected group chiefs, deputies, and team leaders and taught to run the irrigation groups efficiently. They were then required to help the irrigation associations in the distribution of irrigation water, maintenance of irrigation canals, and collection of water fees, and to participate in the training activities conducted by the irrigation associations. JCRR assisted the Provincial Water Conservancy Bureau in meeting the training and supervision expenses and provided guidance to all the twenty-six irrigation associations in executing the program. This has already brought about better maintenance work and has lowered collection and maintenance expenditures in the irrigation associations.

Irrigation Investment Study

In the early days after the retrocession of Taiwan to China, the government concentrated its efforts on the construction of irrigation projects that would produce the quickest results, without giving much consideration to the economic priorities of the different projects. As the number and size of irrigation projects grew but only limited capital was available, it became necessary to make comparative analyses to determine their priorities on the basis of their benefit-cost ratios. Meanwhile, since most of the irrigation associations were not economic-minded in their management and project development, it was considered imperative to improve the situation.

In 1958 the director of the Provincial Water Conservancy Bureau requested the JCRR Rural Economics Division to expand its economic research and statistical activities to include irrigation investment studies and training of the personnel of the Provincial Water Conservancy Bureau and the irrigation associations. The Bureau recruited a number of agricultural economists who, in collaboration with and under the guidance of JCRR specialists, began to conduct agricultural economic surveys, appraise agricultural production in the light of local conditions and national economic policies, evaluate irrigation benefits, and find out the loan repayment capability of farmers so as to establish the economic justification of each project. The irrigation projects that have been carried out after such economic studies include the Tapu Reservoir, the Paiho Reservoir, the Chianan Canal Lining Project, the Erjen Irrigation Project, the Nenkao Irrigation Project, the Houlung Reservoir, and the recently started Tsengwen Reservoir Project. In fact, by now almost all projects, large or small, have been developed by a similar process; their priorities have been determined on the basis of benefit-cost ratio, with due consideration given to their possible overall contributions to the water resources in Taiwan.

Technical Training

To improve the techniques of irrigation and flood control by raising the professional competence or quality of engineers and technicians, selected employees of governmental agencies and public enterprises concerned have been sent abroad for advanced training under the technical assistance program of the United States since 1951. During 1951–1965 fifty persons were trained in the United States, twenty-two in Japan and three in other countries. Most of them have been working for the government agencies or irrigation associations in the field of irrigation, flood control, and water resources development. To improve the skills of young engineers at home, more than ten seminars on irrigation design, flood control, construction methods, and rotational irrigation have been conducted in the past ten years, with a combined enrollment of 646 persons. In addition, to provide an opportunity for the engineers to acquire new knowledge and exchange views and experiences in irrigation and flood control, two symposia were held in 1967, each with 30 participants from various agencies concerned. All these undertakings have helped boost the technical standards of water control and utilization in Taiwan.

Multi-Purpose Project Planning

JCRR takes part in the investigation and planning of almost all major muti-purpose water resources development projects in Taiwan. The projects that have received financial and technical assistance include: the Shihmen Reservoir, the Paiho Reservoir, the Tsengwen Reservoir, the comprehensive planning of the Choshui and Wu

Basins, the irrigation planning of Tachia and Taan Basins, and the Kaoping River Basin integrated development planning.

Research, Demonstration, and Extension

Rotational Irrigation

After years of research conducted by JCRR engineer L. Chow in cooperation with professors of the National Taiwan University and engineers of the Provincial Water Conservancy Bureau, it was found that the conventional irrigation method could be replaced by the so-called rotational irrigation method without affecting production. The data indicated that as much as 20 to 30 per cent of water could be saved in this way. The water so saved could be used to irrigate other areas where water is insufficient or completely lacking. Furthermore, rotational irrigation facilitates plant growth, saves fertilizer, eliminates water disputes, and reduces labor use, especially in drought seasons.

By rotational irrigation, water is supplied in appropriate quantities at the right time and in proper order so that all farmers may get adequate water to irrigate their lands. In order to practice this method, it is necessary to rearrange the existing irrigation distribution systems and improve or install additional irrigation structures. To date, about 77,115 hectares of farm lands in Taiwan have been brought under rotational irrigation. Practically all new irrigation projects have been planned and designed for the extension of the practice.

Moreover, after a land consolidation program was launched by the Provincial Government in 1960, most of the rotational irrigation projects have been carried out in conjunction with the land consolidation projects which have also been supported by JCRR. This has further enhanced the effective utilization of the limited land and water resources on the island.

While rotational irrigation has proved useful and is being widely extended in Taiwan, continued efforts are being made to refine this practice and extend its application to crops other than paddy rice. These efforts may finally lead to the adoption of a unified irrigation practice for both paddy rice and upland crops.

Upland Crop Irrigation

The problem of irrigation for upland crops in southern Taiwan received JCRR attention long ago. But due to water shortage, nothing was done about it until 1961 when an analysis of all the factors affecting crop yield showed clearly the important role played by irrigation. In 1961, the JCRR Irrigation and Engineering Division initiated the first irrigation experimental project on upland crops. It was found that they could be irrigated only during their critical water-demanding periods to get the desired yields. This system made irrigation management less time-consuming and also conserved water. Encouraged by these findings, JCRR in 1963 strengthened the facilities of the Hsuehchia station of the Chianan Irrigation Association and assisted the Provincial Water Conservancy Bureau in establishing five demonstration stations on upland crop irrigation at selected localities all over the island. In 1964 upland crop irrigation entered the demonstration stage. In 1965 and 1966 the experimentation, demonstration, and extension activities were in full swing. Expanded experiments and demonstrations covering many critical points associated with the need for efficient irrigation were extensively carried out at the stations. An area-wide irrigation extension program covering about 5,000 hectares was also started in the Tsaokung Canal irrigation area for promoting winter-crop soybean production. The experience gained in the two years suggested the need for more irrigation extension agents to guide the farmers and further academic research to explain the complicated phenomena observed in field experiments. Therefore, the JCRR 1967 and 1968 programs on upland crop irrigation emphasized basic research in order to learn more about soil-moisture movement and the irrigation requirements of various crops.

Tidal Land Reclamation

The tidal lands on the west coast of Taiwan are sand flats extending from the existing coast line to the line of mean low tide. They are formed by the sediments washed out from river estuaries and deposited along the coast by tide and waves. As the sediment discharge of rivers in Taiwan is rather high, the tidal lands are expanding and growing.[1]

Farmers and fishermen have been making use of the tidal lands on a limited scale for a long time. Paddy fields, fish ponds, and salt beds reclaimed from tidal land can occasionally be seen on the coast.

JCRR first became interested in tidal land development in 1954. The possibility of gaining new agricultural lands by reclaiming on a large scale the tidal flats began to attract the attention of policy-makers. In view of the ever-increasing population pressure on the limited land area, tidal land reclamation was considered one of the measures to boost agricultural production. The first survey of tidal lands initiated by JCRR in 1955 showed 44,000 hectares (53,800 hectares according to the 1961 survey) which, if developed, could be used for agricultural purposes. The reclamation work would involve land leveling and the construction of sea-dikes and irrigation and drainage facilities. JCRR decided on the following steps in undertaking this program:

1. Train the technicians required.

2. Start a pilot reclamation scheme to gain experience.

3. Set up an investigation and planning program to collect basic data, on the basis of which preliminary development plans could be mapped out.

4. Implement reclamation projects first at a few selected places with favorable conditions.

In 1957 the first groups of technical personnel were sent to the Netherlands and Japan for training. A pilot reclamation project was started in 1959 at Hsinchu under the sponsorship of the Vocational Assistance Commission for Retired Servicemen.[2] It was soon followed by another pilot project undertaken by the Taiwan Sugar Corporation at Yunlin. The two projects were successfully completed at the end of 1960.

In March, 1961, the Executive Yuan established the Tidal Land

---

[1] Tidal Land Development Planning Commission, Executive Yuan, "The Tidal Land of Taiwan" (in Chinese) (Taipei: Commission, Nov. 1962).

[2] Vocational Assistance Commission for Retired Servicemen, Executive Yuan, "Development of Hsinchu Tidal Land" (in Chinese) (Taipei: Commission, Aug. 1960).

Development Planning Commission to handle the investigation of all the tidal flats on the west coast. JCRR gave this Commission the necessary technical and financial assistance in setting up meteorological and tidal observation stations, making topographic maps, and conducting soil investigations and irrigation water surveys. Development planning for the tidal lands in Tainan and Chiayi was also completed by this Commission.

The successor to the Executive Yuan's Tidal Land Development Planning Commission is the Taiwan Provincial Government's Tidal Land Development Planning Commission, which was created in 1964 and later renamed the Land Resources Development Commission. JCRR assistance in the fields of investigation and planning was continued despite the changes in organization.

In 1965, the Taiwan Provincial Government started the construction work for the 1,200-hectare Tsengwen polder, which was completed in 1967 and has since been used for fish culture. Two other construction projects were the 227-hectare Hsinchu Tidal Land North Polder and the 1,000-hectare Chiayi Tidal Land Aoku Polder. The former was started in 1964 and completed in 1966 by the Vocational Assistance Commission for Retired Servicemen. The latter was undertaken by the Taiwan Sugar Corporation in 1966 and completed in 1968 by the Land Resources Development Commission. The most recent 462-hectare Wangkung Polder project in Changhua was completed in 1968 by the Land Resources Development Commission. No JCRR financial support was given to any of the construction projects except the Hsinchu tidal land polder.

There have been doubts about the economic value of tidal land reclamation. Due to the complicated nature of the work, the development cost runs very high. But in recent years, the farm land price has also gone up. The reclaimed lands at Hsinchu and Tsengwen polders have all been sold to farmers at cost without difficulty. As Taiwan's agriculture and industry become more highly developed, the value of land will continue to grow.

The total JCRR financial contribution to the tidal land program from 1958 to 1967 was NT$33,793,772, of which NT$25,793,772 was used for investigation and planning and NT$7,880,000 for pilot reclamation of an experimental nature.

Single-Purpose Irrigation Development

Rotational irrigation, canal lining, and ground water development have been implemented as special programs. The main activities in the field of irrigation development in Taiwan have been the construction of single-purpose storage reservoirs, canal and drainage systems, and land reclamation. JCRR has supported the implementation of several significant single-purpose irrigation projects, including three reservoirs, four irrigation canal systems, six pumping plants, and two waste land reclamation projects.

Ground Water Investigation and Development Program[3]

Owing to the uneven seasonal distribution of precipitation, surface irrigation water is generally in short supply in most parts of this island in early spring and late autumn when water is badly needed in paddy fields. Therefore, the area of the first rice crop has to be limited and the production of the second rice crop also often suffers setbacks. Shallow wells had been used as a supplementary source of irrigation water by individual farmers for a long time before the Taiwan Sugar Corporation constructed their own deep wells for sugar cane irrigation in 1949. But the seasonal fluctuation of water table is quite large, causing the shallow wells to yield less in the dry season.

From a geological point of view, the ground water resources in some parts of Taiwan are rich enough to warrant large-scale development. An investigation initiated by JCRR in FY1954 was conducted jointly by the Provincial Water Conservancy Bureau and the Taiwan Sugar Corporation. In 1956, David S. Stoner, a ground water expert of the U.S. Bureau of Reclamation, came to Taiwan at JCRR invitation to review the first part of the investigation report covering the Tachoshui alluvial fan in Yunlin and Changhua counties. Based on his recommendations, the first large-scale ground water development program was started in 1958 in Yunlin and completed in 1961. Altogether 252 deep wells averaging 106 meters in depth and 16 inches

[3] Ground Water Investigation Team, Agricultural Machinery Operation and Maintenance Office of the Taiwan Sugar Corporation, the Taiwan Provincial Water Conservancy Bureau and JCRR, "Investigation Report on Ground Water Resources of the Tachoshui Fan" (Taipei, Nov. 1957).

in diameter were drilled. The total yield of these wells is 238,007 gallons per minute. With the extra water thus made available, a new irrigation area of 15,975 hectares has been created and 9,286 additional hectares of paddy lands have been provided with improved irrigation. The resulting annual increase in rice production is some 55,400 metric tons.

At first, the ground water development program in Yunlin County was handicapped by insufficient funds. Seeing this, JCRR stepped in to give full financial support. A serious spring drought in 1960 caused a general decline in agricultural production except in those areas where some of the wells covered by the project had already been drilled. This fact prompted the Provincial Government to initiate a ground water development project of its own, calling for the construction of 50 deep wells in Changhua County. The project was completed also at the end of 1961, resulting in an annual yield increase of 4,437 metric tons of brown rice.

Another large-scale ground water development project went into operation in FY1962. For this purpose the International Development Association (IDA) extended a loan of US$3,700,000 to the Taiwan Provincial Government. The local currency cost was financed partly by JCRR with a loan and partly by the Provincial Government. This project was completed at the end of 1965 with 418 deep wells drilled, and as a result 59,195 more metric tons of brown rice can be harvested each year.

The beneficial results of the ground water development projects have also led to development efforts by private individuals. Hundreds of wells have been constructed outside the project areas either for irrigation or for fish culture by local farmers.

A ground water investigation project has been in progress for thirteen years. All the plain areas in Taiwan have been surveyed except the Ilan area where irrigation water shortage does not constitute a serious problem. More in-depth studies will be made under this project in the coming years.

The ground water investigation and development program has also proved beneficial to industry. Factories which could not be established for lack of industrial water supply in the past are operating successfully now that ground water is available.

Chianan Canal Lining Program

The rapid population increase and limited water resources call for the economical utilization of water. This can be achieved by canal lining, especially in areas where the available water resources have already been put to maximum use. Canal lining not only saves water, but also has other benefits such as protecting canal banks, controlling weeds, reducing maintenance cost, improving drainage, and increasing the flow velocity and water conveyance capacity.

In the 123,400-hectare Chiayi-Tainan area, where water is so scarce that a three-year crop rotation system has had to be practiced, the total length of the main canals (laterals and sublaterals) is 1,177 kilometers, of which 153 kilometers had been lined by 1959. The seepage loss amounted to 40 or 50 per cent of the water conveyed. It was estimated that the lining of the remaining 1,024 kilometers of canals could reduce the loss to 25 per cent. In this way 100 million cubic meters of water, almost the equivalent of the storage capacity of the existing Wushantou Reservoir, could be saved for the irrigation of rice and other crops in the area. Toward this end, the canal lining project in this area was started in FY1960 under JCRR technical and financial assistance. The whole project was completed in FY1968 at a cost of NT$286,071,000, of which JCRR provided NT$102,200,000 in the form of loans.

Similar but smaller-scale projects have been undertaken by the Kaohsiung and the Changhua Irrigation Associations to prevent percolation losses in waterways that run through sandy soil.

Statistics of JCRR Land and Water Resources Development Program

Land and water resources development has been an important part of the JCRR program since the early years, when it accounted for about 40 per cent of the annual operational budgets.

From 1950 to 1967 inclusive, 476 projects in this category were supported by JCRR, including 232 on irrigation and reclamation, 19 on drainage, 30 on flood control, and 195 on investigation, planning, and experimentation. For these projects JCRR extended about NT$921.7 million and US$2.5 million of grants, and NT$685.8 million of loans. It is estimated that through JCRR assistance about

44,580 hectares of land have been newly irrigated, 467,200 hectares have been provided with supplementary water supply, and 193,000 hectares damaged by the floods of August 7, 1959, have been rehabilitated. The annual increase of production is estimated at about 430,000 metric tons of paddy. Under the flood control program, 108,550 meters of new dikes and 2,300 meters of sea walls have been built, and 175,210 meters of old dikes repaired. Under the Eastern Taiwan Land Development Program, 1,504 hectares of river lands have been reclaimed for agricultural production.

Besides, under the separate Farmers' Organization Program, twelve projects for strengthening the irrigation associations were implemented during the same period. JCRR granted a total of NT$3.4 million for these projects.

The funds spent by JCRR for its land and water resources development program constituted only a small portion of the total investment that had been made in the land and water resources development of Taiwan. JCRR usually required matching or counterpart funds from the government and the people to support the projects.

13. Plant Industry

All technical improvements made in plant industry, livestock, forestry, and fisheries with JCRR assistance to government agencies have to do with better methods and facilities which enable the farmers and fishermen to increase their production. Every improvement is based upon research, experiments, and field demonstration. A few of the most important of these projects will be described.

Crop Breeding

A key factor contributing to increased crop production in the last twenty years in Taiwan has been the successful program for the improvement of crop varieties and seed multiplication. JCRR has given assistance to the agricultural improvement stations of the Taiwan Provincial Department of Agriculture and Forestry and two agricultural colleges for the breeding of crop varieties that grow faster, mature earlier, and have higher yields to fit them into the crowded cropping schedule. A revolutionary cropping system has been established to grow, in addition to two crops of rice, a winter and/or a summer crop, making a total of three or four crops a year on the same piece of irrigated land.

New varieties of rice, sugar cane, pineapples, wheat, soybeans, tobacco, peanuts, sweet potatoes, tea, oranges, mangoes, and watermelons have been developed through introduction, selection, and hybridization, with emphasis on higher yields, better quality, stronger disease resistance, and/or earlier maturity. JCRR has also assisted in the introduction and adaptation of important new crops such as mushrooms, asparagus, hybrid corn, dwarf sorghum, rapeseeds, onions, muskmelons, apples, pears, and a number of new vegetables and green manure crops. But it will suffice to mention, by way of illustration, the breeding of a few outstanding rice varieties.

There are two main groups of rice grown for commercial purposes in Taiwan: the Taiwan native rice, which is a subspecies of the *indica* type; and the *ponlai* rice, which is a subspecies of the *japonica* type.

As compared with *ponlai* rice, the Taiwan native rice is lighter in leaf color, longer and more slender in grain shape, less cohesive after cooking, higher in tillering ability, less responsive to nitrogenous fertilizer, taller, more adaptable to areas where soil fertility is low and the water supply irregular, more susceptible to lodging and shattering, but more resistant to blast disease.

Crossbreeding both among Taiwan native rice and between Taiwan native rice and other introduced *indica* varieties has been carried out at the agricultural experiment stations. Through hybridization and selection, a very promising *indica* variety known as Taichung Native No. 1 was developed and officially registered and released to farmers for cultivation in 1956. This was a cross between two Taiwan native rice varieties, Ti-Chueh-Wu-Chien and Tsai-Yuan-Chung, of which the former was a very short variety with profuse tillerings and the latter a long-stemmed, disease-resistant variety.

Taichung Native 1 has several desirable characteristics. It is short in stature, widely adaptable, and high-yielding. It is not sensitive to photoperiod and can be planted throughout the year, given optimum temperature. Both its response to nitrogenous fertilizer and its yield under fertile conditions are higher than any other *indica* variety or the *ponlai* rice in Taiwan. Because of its short stature, it does not lodge even with heavy application of nitrogenous fertilizer. Its only drawback is that, though resistant to blast, it is susceptible to sheath blight (*Pellicularia sasakii*) and bacterial leaf blight (*Xanthomonas oryzae*) diseases, especially when under conditions of high temperature, high humidity, and close spacing.

Taichung Native 1 is believed to be the first semi-dwarf *indica* rice variety developed by hybridization in the world. Because of its superior quality in so many respects, it has been widely accepted by Taiwan farmers. In 1966, the area planted to it was 77,456 hectares, or 38.2 per cent of the total area planted to Taiwan native rice. It ranks next only to Chianan 8 among all cultivated rice varieties in Taiwan.

In the last few years Taichung Native 1 has been either commonly used as breeding material or directly used by a number of rice-producing countries in Asia for purposes of seed multiplication. In India, for instance, over 250,000 hectares were devoted to the propagation of Taichung Native 1 in 1966. What is more, the successful development of this superior variety has had the effect of overthrowing a hitherto widely held concept that the response of *indica* rice varieties to nitrogenous fertilizer is rather low. On the strength of its excellent showing, Taichung Native 1 has aroused the interest of people in many rice-producing countries and has been chosen by them either for breeding or for propagation. One conspicuous example of the kind is the IR-8 variety developed by the International Rice Research Institute in the Philippines from hybridization between Ti-Chueh-Wu-Chien and Peta, an Indonesian variety.

A *ponlai* variety, Chianung 242, was developed by hybridization from two old *ponlai* varieties, Chianungyu 65 and Chianungyu 123, and was released for general cultivation in 1957. It not only had a high degree of blast resistance, but also gave high yields in many parts of the island, especially in the first crop season. It was a panicle-weight type with high fertilizer response. It was one of the three most popular rice varieties in the late fifties and early sixties, with a total area planted to it of 51,034 hectares in 1962, ranking only below Chianan 8 and Taichung 65. But in later years it was gradually superseded by other newly developed *ponlai* rice varieties, and the area planted to it had dropped to 17,000 hectares by 1968.

Among the old *ponlai* varieties, Taichung 65 was the most popular and had the highest planting area up to 1958. Thereafter, Chianan 8 became the foremost variety. Developed from Taichung 65 and another *ponlai* variety during the Japanese administration of Taiwan, it was, however, selected and released only after World War II. The area planted to it increased from year to year until it reached 128,530 hectares, or 16.6 per cent of the total area planted to paddy rice, in 1966. Though Chianan 8 was susceptible to blast, it was well adapted to the southern and eastern parts of Taiwan and was noted for the stability of its yielding capacity.

The next phase of the *ponlai* rice varietal improvement program was concentrated on the breeding of early-maturing and high-yielding varieties with a fair degree of blast resistance. "Early maturing" meant 100 days from transplanting to harvesting for the first crop, and 80 days for the second crop. The shortening of the period required for maturing was considered necessary to make possible a multiple cropping system under which one winter and/or summer crop could be grown between the two rice crops. Through persistent efforts made by technicians at the various agricultural experiment stations, the first early-maturing *ponlai* rice variety, Taichung 180, was developed and released in 1956. But owing to its low adaptibility, this new variety was soon replaced by a better one, Taichung 186, developed by hybridization from Taichung 65 and Kwantung 5, a Japanese variety. Taichung 186 had a high degree of blast resistance and its yield compared favorably with that of Taichung 65. Still another variety, Taipei 309, developed by hybridization from Kwangfu 401 and Taipei 177, matured one week earlier than Taichung 65 and also yielded better than the latter. It will soon be released to farmers to replace Taichung 65 in the northern parts of Taiwan.

As to high-yielding *ponlai* rice of the *japonica* type with medium blast resistance, there have been developed and released in recent years many new varieties, of which the most promising ones are Hsinchu 56 and Tainan 1 and 5. The first of the trio, developed by hybridization from Tainung 44 and Chianan 2, is highly responsive to nitrogenous fertilizer and very popular with farmers in the Hsinchu area. Tainan 1, developed from Chianan 8, not only yields well but also matures earlier than the parental variety. Tainan 5, also developed from Chianan 8, has the same plant type as Tainan 1, but is more resistant to blast. First released in 1965, it immediately became popular; the acreage planted to this variety was 199,369 hectares in 1968. This was the first new variety to challenge Chianan 8, which had been leading for many years in Taiwan.

One of the old major varieties, Taichung 65, has been losing ground since 1962; the other, Chianan 8, began to do so in 1967.

The *ponlai* rice varieties of Taiwan have attracted the attention of plant breeders in other countries, especially those in Southeast Asia, and have aroused their keen interest. The results of interspecific hybridization show that the *ponlai* rice of Taiwan of the *japonica* type has a higher percentage of fertility than the genuine *japonica* varieties of Japan or Korea when crossed with *indica* rice varieties. Because of this and many other desirable qualities, *ponlai* varieties have been sent to several Asian and nineteen African countries, at their request, for adaptation tests or breeding purposes in the last fifteen years.

The Chinese Government has donated 100,000 metric tons of Chianan 8, Chianung 242, and Tainan 3 to India for seed multiplication. In addition, two metric tons of Tainan 3 have been sent to that country for the same purpose with the assistance of the Rockefeller Foundation. Though the Indian Government is planning to propagate these *ponlai* varieties to increase rice production for their people, it is rather questionable whether the latter will take to them with much enthusiasm.

The objectives of rice breeding in Taiwan are these: to develop *ponlai* varieties that are fairly resistant to such diseases as blast, especially neck blast, yellowing, which is a virus disease transmitted by leaf hoppers, and sheath blight, that mature early enough to fit into the multiple cropping pattern, and that are resistant to lodging in order to tolerate heavy applications of fertilizers and dense spacing; and to develop *indica* varieties that are resistant to the bacterial leaf blight and have good table and milling qualities, but with the same plant type as Taichung Native 1.

As mentioned above, the parental materials used in hybridization for breeding promising *japonica* and *indica* varieties have been the locally bred *ponlai* rice and Japanese rice of the *japonica* type, on the one hand, and *indica* varieties of the Taiwan native rice and introductions from the Chinese mainland, Southeast Asia, and the United States, on the other. The Japanese varieties used in the cross breeding program have been selected for a threefold purpose: to improve the table and milling quality, to shorten the period of growth, and to raise the level of fertilizer response of the existing local rice varieties. But as the present *ponlai* rice varieties in Taiwan have been derived from a temperate-zone ancestry, some new blood from tropical regions is badly needed to boost the low yield of the second rice crop, which has to endure all the severities of a tropical climate. As the American long-grain or medium-grain varieties of the *indica* type are nonlodging, they might improve the Taiwan native rice, if crossed with it.

Seed Multiplication

After the crop breeder has succeeded through experiments and demonstrations in developing improved crop varieties suited to the local conditions, the next step is, of course, the propagation of such varieties through seed multiplication and extension.

In Taiwan as in other countries, the seed multiplication requires the production of foundation seeds, stock seeds, and extension seeds. Under a seed certification system, stock seeds are equivalent to registered seeds, and extension seeds to certified seeds. The system was first applied in 1960 to rice but later extended to nine other crops: sweet potatoes, peanuts, soybeans, wheat, sorghum, barley, rape, corn, and cotton.

Seed Multiplication Problems Peculiar to Taiwan

The small size of Taiwan's farms makes seed multiplication, certification, and distribution more difficult than in countries with larger farms.

In seed multiplication, the great number of small farms has made the technical supervision and field inspection doubly difficult. Even the 15 per cent premium offered to encourage extension seed growers to greater exertions has failed; their farms were so small that any premium they could earn was too small to interest them.

In seed certification, small farms have meant more fields to inspect and more samples to test in the laboratory. This increases the cost of seed production.

In seed distribution, the government bears the cost of collection, transportation, and distribution. A great deal of work must be accomplished in a short time, since the certification procedures must be finished before the next planting season. This is particularly true of the second rice crop which must be planted less than one month after the harvest of the first. Furthermore, some rice farmers will renew their seeds with the certified extension seed for every crop, but others may not take the trouble of going to the farmers' associations for a small quantity of seeds and may go instead to neighboring growers to get the seeds they need. All this has tended to slow down the three-year seed renewal plan mapped out by the government.

JCRR has made continuous efforts since 1961 to overcome these difficulties. By increasing the size of seed farms, doing away with a number of the smaller ones and at the same time establishing large-scale seed farms of rice and peanuts, it has gradually reduced the number of rice seed samples to be tested from 3,147 in 1961 to 1,765 in 1965, and that of peanut samples to be tested from 556 in 1962 to 330 in 1965. The same trend has been evident in the case of other crops and is continuing.

The 15 per cent premium paid to extension seed growers was originally limited to 2,400 kg. per hectare. But as this inducement proved insufficient, the 2,400 kg. per hectare limit has been lifted, and they can now make more profit by selling all the seeds they produce (usually more than 3,500 kg. per hectare), still at a 15 per cent premium. In this way they are encouraged to devote more time to roguing and farm management.

Under a system introduced in 1967, technicians of official and semi-official agricultural organizations such as agricultural vocational schools and farmers' associations are, after having received training in field inspection techniques, qualified to serve as temporary field inspectors during the rush season, in accordance with the rules and regulations governing seed certification. The availability of the services of such licensed field inspectors has contributed to keeping at a high level the quality of the seeds produced by the extension seed growers.

A new approach to the seed multiplication and distribution system has been under trial since 1967. The county government was called upon to divide the total area to be planted to rice in the territory under its jurisdiction into approximately three equal sections. In the first year all the extension seed farms of the county were distributed as evenly as possible throughout the first section. Thus all the rice farmers in the section could conveniently at any time exchange for the required extension seeds since they all had one or more of the extension seed farms near their homes. In the second and third years the same method was followed in the other two sections, so that at the end of the third and last year every rice farmer in the county had his rice seeds renewed according to the government's three-year seed renewal plan.

Vegetable Seed Production

The multiplication and distribution of vegetable seeds in Taiwan are entirely in the hands of private seed growers. The government assists them only by making the foundation seeds of the improved varieties available and by providing them with the technical know-how for producing quality seeds.

Now that vegetable production for both domestic consumption and export has become increasingly important and in consequence annual demands for quality seeds have increased, it is time that seed quality control measures be enforced in order to protect the seed growers themselves, as well as their clients, the farmers.

Since the climate from the central part of Taiwan to the south favors the growth of seeds from the harvesting of the second rice crop in November to the time of the next transplanting in February of the next year, prospects for developing the vegetable seed industry are good. Moreover labor costs are low, and there are a large number of skilled farmers available.

As a first step to boost this promising new industry, JCRR provided financial and technical assistance to the Hsin Sheh Township Farmers' Association in Taichung County for organizing a local seed growers committee to engage in large-scale production of uniform, high-quality radish seeds in 1967. This was intended as a pilot project.

The committee thus organized undertook a number of activities, including cooperative seed production, roguing, and pest control, the improvement of production facilities through a JCRR loan, the offer of a guaranteed minimum seed price to the seed growers to protect them from price fluctuations through another JCRR loan, improved marketing of vegetable seeds through proper packing, and stabilization of the seed market price and the income of the seed growers through the storage of carry-over seeds.

Chemical Fertilizers and Manure

For the more effective use of chemical fertilizers and manure, JCRR has financed research institutes to carry out basic studies on soil fertility, plant nutrition, and their interrelationships. Support has also been given to applied research on the maintenance and improvement of soil fertility, crop responses to added nutrients, and the determination of optimum rates and the timing of fertilizer application for different crops. JCRR has assisted and advised government agencies in training technicians, planning fertilizer requirements, production, and marketing and promoting fertilizer uses. From 1949 to 1967 over 300 such projects were approved by the Joint Commission and implemented by government experiment stations and universities under the supervision of JCRR soil and fertilizer specialists.

In recent years the use of nitrogen, the most important plant food in Taiwan, has been increasing more rapidly than that of either potash or phosphate. In any case, the increased use of all three forms of plant food has had the beneficial effect of raising the per hectare yield as well as the total agricultural production for domestic consumption and export.

The farmers in Taiwan have been using organic manure for centuries. In recent years the annual consumption of compost and farmyard manure has been about ten million metric tons, with a slight increase in farmyard manure and a decrease in compost due mainly to the development of hog-raising and other industries using rice straw as raw material.

An interesting feature in the production of green manure in the last twenty years is the wide fluctuation in the area planted to green manure crops. From 179,000 hectares in 1947, this area first rose to 207,000 hectares in 1952 and then declined to 77,000 hectares in 1966, with a simultaneous increase in the area planted to other economic crops. The crops that depended on green manure in the past are now provided with chemical fertilizers. This change has resulted in a net gain of about 10 per cent in the total cropping area in Taiwan.

Soil Testing and Fertilizer Recommendation

In view of the importance of fertilizers to farm production, commercial fertilizers have been under government control, and rates have been allocated on a county basis determined mainly by the results of regional experiments and demonstrations. However, owing to variations in climatic and soil conditions, experimental data obtained are neither sufficient nor dependable enough to warrant generalizing on the fertility of an area as large as a county. Consequently, the rates allocated are often unsatisfactory in so far as individual farms are concerned. It is quite likely that government control may give way to the free sale of fertilizers in the future in order to assure an adequate supply of fertilizers to all farmers needing them.

At the same time it is necessary to insure that all fertilizer put into the soil will bring maximum returns in crop production. For this purpose, fertilizer recommendations made on an individual farm basis are needed. Toward this end, JCRR decided in 1960 to help the Provincial Department of Agriculture and Forestry develop a soil testing system that would lay a sound basis for fertilizer recommendation.

A long-range project for training soil specialists under the United States Technical Assistance Program was carried out, along with the establishment of a new soil testing laboratory at the Taiwan Agricultural Research Institute, both with JCRR support. The project included the following major activities: (1) establishment and development of suitable chemical methods for use in soil tests; (2) correlation of test values with crop responses to fertilizers under both field and greenhouse conditions; and (3) conducting a fertility survey of cropland on an island-wide scale.

For the fertility survey, 78,635 soil samples from 786,350 hectares (about 88 per cent of the total cropland) were collected and analyzed for soil texture, pH, organic matter content, and available phosphorus and potassium. Results of the tests thus obtained can, if plotted on 1:100,000 scale maps, serve as useful reference material in making fertilizer allocation and recommendations.

Detailed Soil Survey of Cropland

Initiated by JCRR and jointly sponsored by the Taiwan Provincial Department of Agriculture and Forestry and the Provincial Chunghsing University in 1963, this survey called for the classification and mapping of all the agricultural soils of Taiwan within seven years. By the end of 1969 detailed soil maps on a 1:25,000 scale had been made of 500,000 hectares (about 55 per cent) of the cultivated land of Taiwan. Soil survey reports and maps for Changhua and Tainan counties have been published and those for Chiayi, Pingtung, and Kaohsiung counties are being produced.

Plant Protection

Starting from a humble beginning in 1950, the plant protection service in Taiwan has kept pace with the rapid advances in crop improvement made over the years. Several newly developed agricultural industries in the province owe their present state of development and even their very existence to effective pest control. Sometimes, a sudden outbreak of unfamiliar pests has threatened to destroy an important plant industry, and experimental workers have been called upon to improvise methods of control.

Not many years ago the farmers in Taiwan suffered heavy losses from pests. But the introduction of modern agricultural chemicals in the early fifties paved the way for crop improvements other than pest control. Today, pest control has become a standard procedure in crop production and promises to have a profound influence on the further development of agriculture in Taiwan.

Insect Control

One outstanding example of insect control is illustrated by the case of the pineapple mealy bug, which was first brought into Taiwan on pineapple seedlings in 1935 or earlier and became a limiting factor in pineapple production in the southern parts of the island. The insect, which was found to be the cause of the "pineapple wilt," rendered large areas unsuitable for pineapple cultivation during World War II and immediately thereafter.

Through research, an effective control of the mealy bug-induced wilt was found. Not only have large areas been saved for pineapple cultivation, but the practice of close planting of pineapples has also been made possible. The overall result has been a more than twofold increase in the per hectare yield, from 9,698 kg. in 1952 to 22,501 kg. in 1966, together with a significant rise in the annual volume of canned pineapple export.

There are many other examples of the progressive use of pesticides which have contributed greatly to agricultural production in Taiwan.

To protect both the producers and users of pesticides, regulations governing the registration, standardization, and sale of pesticides have been promulgated and enforced since 1959. A physical-chemical laboratory and a biological testing laboratory have been set up to check the quality of the products on the market.

Disease Control

Before effective measures were developed for its control, rice blast had caused much damage to crop production in Taiwan, with an estimated loss of 5 per cent of the total rice production in 1952. Though the threat is still there, serious occurrences of the disease have been rare in recent years as a result of the intensive control program implemented with JCRR assistance.

All new rice varieties for commercial growth in Taiwan are released only after they have been successfully tested at the disease testing nurseries. Any variety that lacks blast resistance is not released. It should be noted, however, that this type of testing does not duplicate the screening test for the selection of disease-resistant lines from their early generations, which is usually carried out by the breeding stations themselves.

Of all the new rice varieties put to nursery tests and recommended for commercial growth, Chianung 242 of the *japonica* type and Taichung Native 1 of the *indica* type have the highest level of blast resistance. The area planted to Taichung Native 1 amounted to 77,456 hectares, or 38.2 per cent of the total area planted to *indica* type varieties, in 1966. The area planted to Chianung 242 reached its peak in 1962 with 51,034 hectares, which was about 10.2 per cent of the total area planted to *japonica* varieties in that year. However, varieties that are less blast resistant but have other desirable characteristics have been gaining favor with the farmers, and the area planted to Chianung 242 has gradually declined ever since. This new development has been made possible only because effective chemical controls, such as field fungicides, are now available for use against rice blast.

The measures recommended by JCRR to control the outbreak in the early sixties of a new rice disease in certain parts of southern Taiwan show how activities in plant pathology contribute to increased crop production. At first incorrectly identified as "rice suffocating," the disease broke out during the second rice crop in the Pingtung area in 1960; occurring in different degrees of severity, it affected some 13,770 hectares of paddy field and inflicted a loss of 18,281 metric tons of brown rice in harvest. The disease continued in 1961, affecting 14,986 hectares, and reached a peak in the following year, when it affected 24,945 hectares.

A JCRR plant pathologist, Dr. Chiu Ren-jong, observed in 1962 that healthy and diseased plants often existed side by side in the rice field. Further experiments undertaken by him and his associates in 1963 gave evidence that a disease characterized by yellowing of the lower leaves and reduced tillering such as that found in Pingtung and many other areas could be transmitted by the leafhopper, *Nephotettix apicalis,* a common rice pest in Taiwan. Later work by the Provincial Chunghsing University produced evidence indicating that a new virus was the real cause of the disease in most of the rice-growing areas where the so-called rice suffocating had been previously reported.[1]

Attacking the problems from two different angles proved to be highly rewarding. It has been found that the disease that broke out in some parts of southern Taiwan in the early sixties was not one but two new diseases of different nature. The recognization of the true nature of the disease and the prompt steps taken for its control have led to a gradual decline in the area affected by it and in its severity, wherever it occurred, in recent years.

[1] Chiu Ren-jong *et al.,* "Transmission of Transitory Yellowing Virus of Rice by Two Leafhoppers," *Phytopathology, the Official Journal of the American Phytopathological Society,* Vol. 58, No. 6 (1968), pp. 740–745.

Systems of Multiple Cropping

One of the special features of agricultural development in postwar Taiwan—the adoption of systems of multiple croppings, that is, the production of three or four crops a year on the same piece of land—has been described in the writer's *Agricultural Development on Taiwan since World War II*,[2] to which the reader can readily refer. Therefore, only certain new developments in multiple cropping will be discussed here.

Factors Contributing to Multiple Cropping in Taiwan

The successful implementation of multiple cropping in Taiwan has, in a large measure, depended on the availability of ample irrigation facilities, early maturing varieties, cultural improvements, the rational application of fertilizers, effective control of crop pests and diseases, and the adoption of farm machines. Moreover, the successful land reform projects and fair prices for most crops have also acted as powerful incentives to the farmers to adopt systems of multiple cropping which promise maximum returns from the land.

The importance of irrigation to the practice of multiple cropping has been clearly demonstrated in the Tsaokung Canal district in southern Taiwan, where the irrigated area covers fourteen townships with a total of 11,255 hectares of farm land. In the past the farmers in the district grew two crops of rice a year and let most land lie fallow in winter, mainly because irrigation was stopped after the second harvest. During the winter season only about 3,000 hectares of the paddy land were planted, mostly to native soybean varieties that could survive without irrigation but gave very low yields. These yields were in sharp contrast to the high ones of the irrigated soybean fields in neighboring Pingtung County.

To promote better land use for maximum crop production in this district, a comprehensive improvement project was initiated by JCRR specialists. After a thorough discussion with the crop experts of the Kaohsiung District Agricultural Improvement Station in 1964, the

[2] T. H. Shen, *Agricultural Development on Taiwan since World War II* (Ithaca, N.Y.: Cornell University Press, 1964), pp. 155–164.

project was put into operation in the following year with the support of the Kaohsiung Irrigation Association and the local governments concerned.

The project had a threefold purpose. First, it would readjust the prevailing irrigation schedule to make irrigation water available to winter crops, especially soybeans planted right after the second rice crop. Second, it would introduce early-maturing rice varieties to make possible the growing of an additional soybean crop in mid-October. Lastly, in order to boost the yields of both the rice and the winter crop, it would give large-scale integrated demonstrations of improved varieties, cultural practices, and collective pest control.

Under the new system of cropping, yields of rice and soybeans increased, and so as a result did the farmers' incomes, as shown in Table 13-1.

Table 13-1. Crop yields and incomes in the Tsaokung Canal district in 1964, under the old cropping system, and in 1965–1966, under the new system

Year	Crop	Variety	Days of growth*	Unit yield	Net income† NT$/ha.	Index	Acreage	Total net income†
				(kg./ha.)			(ha.)	(NT$1,000)
1964	2nd rice crop	Native variety	130	3,094	4,350	100	1,500	6,525
1965	2nd rice crop	Improved *ponlai*	97	3,952	7,636	—	1,477	11,278
	Soy-beans	Improved varieties	90–95	1,371	3,591	—	1,401	5,031
Total					11,227	258	—	16,309
1966	2nd rice crop	Improved *ponlai*	97	4,524	9,827	—	1,516	14,898
	Soy-beans	Improved varieties	90–95	2,154	7,897	—	1,487	11,743
Total					17,724	407	—	26,641

* From transplanting to harvesting.
† The net income is calculated on the following basis: (1) unit price NT$3.83 per kg. of paddy rice and NT$5.50 per kg. of soybeans; (2) cost of production NT$7,500 per ha. of paddy rice and NT$3,950 per ha. of soybeans.

With such advantages in its favor, the new cropping system has been readily adopted by the local farmers. The area devoted to winter cropping was over 6,000 hectares in 1967 and extended to 7,566 hectares in 1968.

Farm Mechanization

The farm mechanization program in the last fifteen years has resulted not only in an increase in the number of farm machines, but also in the development of many kinds of processing and cultural equipment including corn shellers, jute decorticators, sweet potato diggers, grain dryers, hand soybean planters, peanut threshers, tea cutters, and rice transplanters. Table 13-2 shows the steady increase in the number of power tillers since 1954, the first year when a few imported from abroad were used on a trial basis; the number of locally-made tillers has also increased.

Table 13-2. Increase of power tillers in Taiwan, 1954–1968

Year	Total number power tillers	Imported power tillers (%)	Locally-made power tillers (%)
1954	7	100	0
1957	180	93	7
1960	3,708	54	46
1963	9,079	48	52
1966	14,272	32	68
1967	17,240	26	74
1968	21,153	21	79

The first seven power tillers were garden tractors imported from the United States. As they proved useful only on dry land, not in paddy fields, two Japanese-made ones were imported in 1955. One of them, a 2.5 horsepower gasoline-engine tiller, was found after repeated tests at the agricultural experiment stations to be well adapted to the local farming conditions because of its light weight, easy maneuverability, and reasonable price. It could also be used for various farming operations such as tillage, transportation, and cultivation.

In the next year JCRR imported some more Merry Tillers, as they were called, for testing and demonstration at the agricultural stations. During the demonstrations the local farmers showed an intense interest in the machine. The demand for it was so strong that some enterprising local machinery manufacturers began to produce it by copying the imported ones. In just two years the number of small farm machinery manufacturers had gone up to twenty-two. Consequently, many kinds of power tillers were produced and appeared on the market. Soon there were as many as sixteen brands of imported power tillers for the farmers to choose from.

In the late fifties, experiments were conducted with both the imported and locally-produced power tillers on various soils and crops at the agricultural improvement stations. At the same time, training classes were held, regulations governing fuel supply drawn up and promulgated, and inspection systems planned with JCRR assistance.

While the number of power tillers increased from year to year, the number of draft animals declined. From 1961 to 1966 there was an average decrease of 10,000 head a year. As the farmers became increasingly aware of the benefits of farm machines, their demand for them encouraged both the introduction of new machines from abroad and the local production of power pumps, mist blowers, grain dryers, power threshers, and processing machines.

To disseminate the technical know-how needed to use farm machinery, especially power tillers, among the basic-level agricultural extension workers and the farmers, JCRR assisted the agricultural agencies in conducting training classes and demonstration activities. In the first five years of power tiller extension, 943 persons were trained to carry out tests and demonstrations and to train others in the use, care, and simple repair of power tillers and other agricultural machines. They included basic-level extension workers from county governments and township farmers' associations and skilled workers in charge of farm machine repair and maintenance. Some of these men, in turn, taught 781 farmers to operate and maintain power tillers and make simple repairs.

Since the first two major farm machinery firms, the China Agricultural Machinery Company and the New Taiwan Agricultural Machinery Company were established in 1961, they have taken upon themselves the responsibility of training extension and service personnel as well as farmers for the benefit of all parties concerned. Simultaneously, JCRR has continued to help the Provincial Department of Agriculture and Forestry and the local governments hold short-term training classes for farmers to enable them to learn more about farm machines and also to improve their skill in operating and maintaining power tillers.

Farm machinery demonstrations, field days, and mechanical farming contests are often held by the agricultural agencies concerned to arouse the farmers' interest in farm machinery.

Promotion of Research

The imported power tillers and other farm machines usually need modifications to fully meet the local agricultural demands. Since 1957 JCRR has helped the agricultural research and improvement stations train their farm-machinery research personnel; it has also subsidized the construction of workshops to improve the performance of power tillers by designing and developing many kinds of attachments and farm implements. When machines have to be produced locally, the materials or designs are often subject to change. For this reason, JCRR has assisted the local farm machinery manufacturers in setting up their own research departments staffed with qualified engineers to make improvements on the power tillers and develop new machinery and implements. But as the engineers working for the local farm machinery manufacturers usually lack field experience and as the government agencies are short of qualified research engineers, a JCRR farm machinery specialist has worked out a mutual cooperation program whereby close liaison is maintained between government research workers and local manufacturers. The implementation of the program has speeded up the research work on farm machinery and implements in recent years.

Production and Testing of Power Tillers

In the first few years after the introduction of power tillers from abroad, some inexperienced small manufacturers tried blindly to copy foreign models; most of them failed to produce serviceable machines and were soon forced out of business by cutthroat competition. In order to benefit by the manufacturing techniques, experience, and financial backing of the advanced countries, JCRR supported the cooperation between Taiwan machinery manufacturers and Japanese farm machinery companies in 1961 in establishing the China Agricultural Machinery Company and the New Taiwan Agricultural Machinery Company for the production of power tillers and other farm machinery in Taiwan. The two companies, which have progressed steadily, are now able to produce serviceable power tillers. Thus the need for imported machines has been greatly reduced, as shown in Table 13-2.

To insure the quality of power tillers turned out by the big and small plants, a power tiller standard was set by the responsible government agency and the testing facilities for conducting inspection work in the College of Agriculture of the National Taiwan University were strengthened with JCRR assistance.

Establishment of Farm Mechanization Promotion Centers

As farm mechanization is bound to gain popularity in the years to come, JCRR has, since 1965, rendered necessary assistance to the Provincial Department of Agriculture and Forestry and the local governments for the establishment of township farm mechanization promotion centers in selected townships. The centers' main functions are: (1) to maintain and repair farm machines owned by individual farmers; (2) to fully utilize farm machines, especially power tillers, by making their services available to other farmers through arrangements with the center; (3) to assist any farmer who wishes to buy farm machines but lacks funds by helping him complete the bank loan procedures; (4) to assist the research institutions in field testing and demonstration of newly developed farm machines; and (5) to carry out a farmers' training program in each of the townships where a promotion center has been set up.

Fruits and Vegetables

Fruits and vegetables, important both for home consumption and as earners of foreign exchange, have been given special attention.

With the large influx of people from the mainland to Taiwan that began in 1949, vegetables and fruits as well as other kinds of food were in short supply. As the Chinese people consume more vegetables than meat, the shortage was keenly felt. In 1950, JCRR began helping the experiment stations of the Taiwan Provincial Department of Agriculture and Forestry to improve and increase the production of vegetables and fruits. Horticulturists introduced many varieties from abroad and handed them over to the experiment stations to be grown and tested. After a few years of experimentation and testing, the better ones were selected for extension as shown in Table 13-3. In this way

Table 13-3. Introduction and extension of vegetables in Taiwan since 1948

Vegetable crop	Year of Introduction	Year of Extension	Introduced by
Arrowhead	1957	1958	Taipei DAIS*
Asparagus	1955	1964	JCRR
Asparagus let tuce	1949	1956	Fengshan THES†
Broccoli	1955	1957	JCRR
Brussels sprouts	1956	1960	JCRR
Chantenay carrots	1956	1957	JCRR
Day lilies			Traders
Kohlrabi	1956	1957	JCRR
Mushrooms	1953	1956	JCRR, TARI‡
Muskmelons	1956	1957	JCRR
Nodese wax gourds	1948, 1960	1961	Traders, Fengshan THES†
Norin No. 1 potatoes	1956	1957	TARI,‡ JCRR
Onions	1951	1954	JCRR, Taipei DAIS*
Szechuan mustard	1950	1956	TARI‡
Water chestnuts	1953	1954	TARI‡

* DAIS = District Agricultural Improvement Station
† THES = Tropical Horticultural Experiment Station
‡ TARI = Taiwan Agricultural Research Institute

the production of vegetables and fruits was gradually stepped up and their quality much improved.

Large quantities of pineapples, bananas, citrus fruits, mushrooms, onions, tomatoes, and asparagus are being exported to many parts of the world. Most of them enjoy a good reputation in the world market.

Improvements in Vegetable and Fruit Production

New varieties and methods introduced and demonstrated by the various agricultural agencies with JCRR support have resulted in great improvements in vegetable and fruit production.

Early in 1950, JCRR introduced the artificial culture of mushrooms. They proved a highly successful crop for both domestic consumption and export. From late October to early April the weather in Taiwan favors their growth. Farmers found that with the help of their families they could grow mushrooms as a sideline. The crop could be grown in simple structures with very little cash outlay for spawn and fertilizers.

A small quantity of mushroom spawn was introduced from the United States by the Taiwan Agricultural Research Institute in 1953. The next year JCRR financed the Institute with the equivalent of only US$594 in local currency to carry out an experiment using easily available materials for growing mushrooms. After a number of tests, it was found that mushrooms would thrive best if a synthetic compost consisting of rice straw and chemical fertilizers such as urea and superphosphate was applied.

The extension of mushroom cultivation began in earnest in 1956. It was estimated that by October, 1963, some 50,000 farm families were engaged in mushroom growing. Approximately 25,000 people—including those working at mushroom collection centers and the canneries, local wholesalers, retailers, and vendors—were also in the mushroom business.

Between 1962 and 1968 both the production and export value of mushrooms more than tripled, as shown in Table 13-4. The farmers'

Table 13-4. Mushroom production and export in Taiwan, 1962–1968

Year	Production (M.T.)	Export Amount (cases)*	Export Value (US$)
1962	15,680	709,000	8,508,000
1963	38,639	1,378,000	16,148,000
1964	22,718	1,172,000	15,605,000
1965	32,430	1,606,000	20,481,000
1966	38,454	1,845,000	25,085,000
1967	50,181	2,303,000	32,228,000
1968	52,401	2,310,000	32,320,000

* 1 standard case = 48 lbs. (gross weight), or 18.5 kg. (net weight)

profit from mushroom growing averages about 10 to 20 per cent of the production cost, including family labor, building materials, etc.

Outstanding Problems

Annual typhoons are a serious threat to vegetables and fruits in Taiwan; in spite of the windbreaks and drainage systems that have been widely adopted to break the force of the violent winds and prevent flooding, the tropical storms and torrential downpour can still cause heavy damage. This is especially true in low land areas such as those in the vicinity of Taipei City, which often suffers shortages of vegetables and high prices in the wake of typhoons. Human efforts can only minimize the damage.

Fruit growers face the problem of soil erosion on slope land where most of the fruit orchards are located. In the absence of proper soil conservation practices, erosion has caused the loss of top soil, reduced soil fertility, and resulted in decreased production and greater dangers of flooding. Though some efforts have been made to preserve the soil, limited manpower and financial support have failed to bring about a general enforcement of soil conservation practices. As only a small percentage of the fruit growers have adopted proper conservation measures on slope lands, the matter is one of serious concern.

A third problem affects both vegetable and fruit growers—there are no large seed companies with well-established reputations which supply reliable certified seeds, and there are no laws and regulations governing seed transactions and prohibiting the sale of poor and low quality seeds. Most seeds that are presently available on the local market are of average quality and not entirely dependable. Though there are some selected and certified seeds in the absence of any legal restrictions, they cannot compete with the inferior, uncertified seeds owing to their higher cost. Under such circumstances, the production of vegetables and fruits suffers in both quantity and quality. Moreover, since there are not enough seeds of selected varieties for general use, production of quality products is limited.

Mainly because of poor soil fertility, improper application of fertilizers, inappropriate spacing between plants, and inadequate disease and pest control, the unit yields of fruits in Taiwan are lower than the world average. Even in the different sections of Taiwan, yields vary greatly. Thus, the per hectare yield of bananas is only 6,000 kg. in central Taiwan as compared with 20,000 kg. in the south. It is believed that improvements can be made in most, if not in all, cases. As it is, however, only a few citrus fruit orchards are up to the required high standards in both management and production practices.

The quality of several major horticultural products needs to be raised. Up to the present, only the pineapples used for canning meet the international quality standards and can stand foreign competition.

Seasonal disparity between the supply and demand of some fruits and most vegetables presents another difficult problem to all parties concerned. For instance, the peak market demand for bananas in Japan occurs between August and March, a period during which banana production in Taiwan is the lowest. On the other hand, when banana demand in Japan is low during the summer months owing to the abundant supply of other fruits, banana production in Taiwan is at its peak. Inevitably, this has resulted in the loss of part of the Japanese market to banana growers in Taiwan, and in low prices.

Vegetable production is at its peak between October and the following April because of rotational cropping and climatic conditions. After May, the supply of fresh vegetables gradually declines until there is a real shortage in July, August, and September. This is, therefore, another pressing problem waiting to be solved.

Lastly, better coordination between production and marketing has yet to be achieved. There have been some improvements in this regard in the case of bananas, citrus, and pineapples, with the fruit producers' cooperatives acting as coordinators between the growers and the exporters, and looking after the interests of the former. The maintenance of a uniform price for pineapples set by the government has minimized conflicts of interest between the growers and the canneries. But occasional adjustments in the price have often induced the farmers to pick the fruits either too early or too late in their attempt to fetch a better price at the expense of fruit quality. Proper measures are needed to establish a sound relationship between production and marketing.

A Horticultural Crop Survey Team to Other Countries

In view of the immense success achieved through the introduction of hundreds of varieties from abroad for use by the local experiment stations, a five-man horticultural crop survey team headed by a JCRR specialist, Mr. Chi-Lin Luh made an extensive tour of many countries during 1967 to collect specimens of crop varieties for further experimentation. The trip took them to the Philippines, Australia, the Fiji Islands, American Samoa, Hawaii, Mexico, Guatemala, Honduras, Costa Rica, Panama, Peru, Chile, Argentina, Brazil, Trinidad, Puerto Rico, Jamaica, and Pennsylvania and California in the United States. From these places the team sent back by air fifteen shipments consisting of planting materials for 891 varieties of 117 kinds of crops.

Most Latin American countries are rich in bean varieties. Those collected from there by the survey team should be valuable for increasing the plant protein supply in Taiwan and possibly in other tropical countries as well. JCRR is closely watching the development of seeds, bulbs, and other planting materials sent home by the team. Should some of the varieties show promising results, a new plan for multiplication and regional testing will immediately follow. There is every hope that good new varieties and crops developed from this collection will enrich the agricultural resources of Taiwan.

Integrated Improved Techniques and Joint Farming Operations

As industrialization has advanced in Taiwan, more rural people have been absorbed into industry and other nonagricultural work; thus the percentage of agricultural population to the total population has been steadily declining. But owing to the population growth, the absolute number of the agricultural population has been increasing. Between 1953 and 1968, the percentage of agricultural population to the total population dropped from 52 to 44, but its absolute number increased from 4,382,000 to 5,999,000. In the same period, the average size of the farm has dropped from 1.24 to 0.92 hectare. There is little hope of bettering the situation.

To offset the obvious disadvantage of the small family farm, which was only 1.1 hectares in size in 1960, JCRR, in consultation with the Taiwan Provincial Department of Agriculture and Forestry, has initiated a method whereby these farms can achieve the productive efficiency and the advantages of large-scale operations. The work is divided into three stages.

The First Stage

A JCRR rice specialist, Cheng-Hwa Huang, assisted the PDAF in starting in 1963 a demonstration of the simultaneous application of improved cultural techniques. The package approach involved the use of superior rice varieties, the raising of healthy seedlings, their transplanting at optimum spacing with the optimum number of seedlings per hill, the proper application of fertilizers, intensified disease and insect pest control, better irrigation and drainage, and more careful and thorough weeding. The demonstration, which started with the fall rice crop of 1963 and the spring crop of 1964 at four localities, called for joint operations by farm families on demonstration units of approximately ten hectares each. About ten farm families pooled their manpower and material resources to work on one such unit, with JCRR and the PDAF exercising technical supervision and furnishing the necessary fertilizers, pesticide sprayers, and power tillers. The demonstration was continued during the next two rice crops in the fall of 1964 and the spring of 1965, but this time 15 other localities were chosen and the size of the operating units was reduced from ten to five hectares each for easier management.

The farmers in each unit planted a given variety on the same days and used the same cultural methods so that the rice plants throughout the unit would simultaneously reach the same stage of growth. This facilitated the simultaneous application of fertilizers, pesticides, and irrigation water as well as simultaneous weeding and harvesting. On the average, the rice yields from the demonstration units were 34 per cent higher than those from nearby fields in 1963–1965, and the net income 48 to 50 per cent higher. The adoption of only one improved measure, whether it was an improved variety, better fertilization, or pest control, had increased the yield by less than 10 per cent according to previous records.

The 34 per cent increase in rice yields was the result of a combination of many favorable factors: the small area (5 hectares) of demonstration farms, selection of the best varieties, adequate financial support to provide the optimum and timely application of fertilizers and pesticides, the liberal use of labor, and, last but not least, close and constant supervision by rice specialists. If any one of these factors had been lacking, the results could not have been so outstanding, as the later extension program showed.

The Second Stage

Since these demonstrations proved successful, JCRR extension specialists, in cooperation with the PDAF and the farmers' associations, in 1964 started a program on integrated improved techniques through joint rice farming. This time the emphasis was shifted to the organizational aspects; improved cultural techniques were used experimentally on larger operating units. Every unit has contained about 15 hectares and has been cultivated by twenty to thirty farm families, depending on the size of the unit. JCRR has given technical assistance and has made a small grant of NT$7,000 to each of the units for the purchase in the first year of a power sprayer.

The operating unit elects its own leader. An inventory is taken of the farm area, facilities, farm tools and equipment, land fertility, and available labor force of each of the farm families belonging to the unit. With this information, and with the help of agricultural extension agents, the leader draws up a work plan for his unit by incorporating in it all the improved cultural methods that have proved successful.

The members of an operating unit are divided into several working subgroups, to each of which is assigned some specific task such as land preparation, transplanting, fertilizer application, irrigation, weeding, pesticide spraying, and harvesting. Every subgroup works under the supervision and guidance of a leader chosen on the strength of his farming skills from among the members of the group. This has proved to be far more effective than the traditional hand cultivation with each farm family functioning as an isolated farming unit.

Convinced of the benefits of the joint operating units, many farmers are eager to participate in them for a variety of reasons. Those without sufficient labor to cultivate their own fields want to avail themselves of the surplus labor of their neighbors. On the other hand, farm families with too much labor feel that participation will automatically give their surplus labor full employment. The same is true of farmers with farm machines such as power tillers and pesticide sprayers, and of those without them. Both groups are interested in joining, but for opposite reasons.

Farmers in Taiwan are accustomed to exchanging labor, hiring draft animals and labor to prepare fields for planting, etc. So the joint operation to promote adoption of improved practices is readily accepted. The purpose is to upgrade the level of technology on all the land in the extension area, including that of small farmers who otherwise might not think the improved techniques would be profitable.

It must be pointed out, however, that the system of joint farming in Taiwan is different from that of the cooperative farm tried out by the Communists on the Chinese mainland. The difference lies in the fact that each farm family taking part in the joint farming still has exclusive rights to the land, tools and implements, and the produce from its own fields; as a result the family has every incentive to increase production. All members of an operating unit work jointly and are paid wages for all labor done at a scale set by the farmers themselves beforehand. As plowing, transplantings, pest control, and harvesting call for either some skill or substantial physical exertions, the wage scales for these jobs are higher than those for less arduous tasks such as applying fertilizer, weeding, and irrigation. With the wage scales thus set, the farmers best suited for any given kind of work are assigned to do it at the appropriate time. After the completion of this work, the number of man-days spent on the 15 hectares of the operating unit are multiplied by the daily wage to arrive at the total wage for the specific task. By dividing the total wage by 15, the per hectare wage is calculated. Following that, the wage each farmer has to pay for the work done on his farm, whatever the size may be, is calculated.

The next step is to calculate the wage each farmer and his dependents should receive for the labor they have contributed. By multiplying the number of man-days they have contributed to each task by the daily wage set for it, and subtracting from the total the farmer's payment to others for the work done on his own farm, the net wage payable to him and his dependents is determined. However, if the total wage he and his dependents should receive is less than the total wage he should pay to others, he is called upon to make up the difference.

The wage rates paid to farmers for the different kinds of work done in joint farming operations, which are much the same as those prevailing in the area, are shown in Table 13-5. As the table indicates, the daily wage for plowing is highest (NT$120 as compared to NT$40 to NT$60 for other work), in part because it demands the most physical exertion and in part because a plougher must use his own draft animal.

It is the general practice in Taiwan for an employer on a farm to provide his labor force four or five meals a day, plus such incidentals as tobacco and tea in addition to cash wages. But since the farmers taking part in the joint farming operations work for themselves as well as for other members of the unit, no extra meals are provided. However, despite the lack of meals and the fact that wages are not higher than standard rates, the farmers benefit greatly from the increased farm labor productivity in the residual returns to land, management, and capital.

A study of Table 13-5 shows three important facts respecting labor productivity measured in terms of the total production per unit of farm labor input. First, production per hectare, or yield (kg/ha), is higher on the joint operation plots than on the corresponding check plots. Second, the total number of man-days per hectare (man-day/ha) is less on the joint operation plots than on the corresponding check plots. Third, labor productivity (production in kg./man-day) is higher on the joint operation plots than on the corresponding check plots. On the average, the joint operation plots have from 25 to 30 per cent higher productivity than the corresponding check plots practicing individual operations.

Table 13-5 also shows that joint operations tend to save labor input in all kinds of farm work in rice cultivation, but the degree of labor saving differs with the type of work. The larger units make it possible to use power tillers, mist sprayers, and other mechanical devices, making the degree of labor saving very high in nursery management, irrigation, and drainage. It is much lower in transplanting, weeding, and harvesting, mainly because the higher density planting practiced in the joint operations requires more labor. This would indicate that selective employment of measures is more important than a uniform approach to effect labor productivity improvement.

To make it economically significant, the labor saved through joint operations should have alternative employment opportunities. Efforts

Table 13-5. Labor productivity on joint-operation plots and on check plots, and wage rates paid for farm work*

Labor and productivity	1st rice crop, 1966 (man-days/ha.)		2nd rice crop, 1966 (man-days/ha.)		Wages (NT$/day)
	Joint opera-tion	Check plot	Joint opera-tion	Check plot	
I. Man-day input per ha.					
1. Nursery management	2.24	3.42	2.19	3.35	60
2. Ploughing of paddy fields (including services of draft animal)	19.33	20.57	18.55	19.42	120
3. Transplanting	9.35	9.49	9.60	9.80	60
4. Weeding and fertilizer application	20.63	22.04	18.96	20.27	50
5. Irrigation and drainage	5.48	8.50	6.06	8.32	40
6. Pesticide spraying	5.65	6.46	5.73	6.60	60
7. Harvesting	17.49	18.30	17.83	18.73	60
Total number of man-days per hectare	80.17	88.78	78.92	86.49	
II. Total production of paddy rice kg./ha.	4,708	4,004	5,223	4,561	
III. Labor productivity kg./man-day	58.71	45.10	66.17	52.73	
Index	130	100	125	100	

* The figures in this table are based on the data collected from 26 joint operation and corresponding check plots in Taichung County.

should be made to expand the intensive cultivation of crops other than rice and to develop industries in rural areas. Participation in training and educational programs would give the farmers more scientific information, and the joint farming operation would facilitate the dissemination of such information.

In 1964 one hundred joint farming units were established in 86 townships. Because they achieved such significant increases in yield, about an equal number of joint operating units was added in each of the two following years, thus boosting the total number to 296 by the end of 1966.

The advantages enjoyed by farmers taking part in joint farming operations with integrated improved techniques are obvious. The per hectare yield from farms under these operations has been 11 to 15 per cent higher than that from the check fields during the three years under observation. However, as a result of the gradual adoption of improved techniques on the neighboring check fields, their rice yield also increased, and consequently, the per hectare yield index for farms under joint operation was smaller in the third than in the first year.

In this connection, it must be pointed out that the per hectare yield achieved on the farms under joint operation during the extension stage did not equal the 34 per cent higher yield achieved on the demonstration farms because some of the favorable factors present during the demonstration stage were absent at the time of extension.

When the system of joint operations with integrated improved practices was being extended, the farmers did not necessarily use the best new rice varieties, nor did they receive any financial assistance to enable them to apply the optimum amount of fertilizers and pesticides and the necessary labor. Moreover, the needed fertilizers and pesticides might not have been available at the right moments. All these less favorable factors, together with the supervision by extension agents instead of JCRR and Provincial Government rice experts, caused a less impressive yield increase than that achieved during the demonstration.

However, on the recommendation of extension advisers, the farms under joint operation applied 4 to 12 per cent more fertilizers and 52 to 85 per cent more pesticides than the neighboring check fields. By adopting the recommended technique of thin sowing, they also used 5 to 10 per cent less seed of improved varieties of rice.

It may be added that the services of specialists and extension agents are likely to be more effective with larger units than with individual farms. The integrated improved techniques have reduced the unit cost of production by 10 to 20 per cent and increased net income per hectare by 21 to 31 per cent.

The Third Stage

The third and last stage of integrated improved techniques through joint farming operations was started with the second rice crop of 1967 at five townships, covering nearly three thousand hectares per crop.

All five townships will continue the integrated package program for three years in succession. Two other townships were added in 1968. It is expected that the average rice yield at these selected townships will be significantly increased as a result of the successful implementation of the program.

The farmers in the program are divided by villages into operating teams which, in turn, are subdivided into working units. Each unit consists of 30 to 40 farm families who work together to cultivate a 20-hectare plot. Each unit elects its own leader; local extension workers serve as technical advisers under the guidance of technicians from agricultural stations.

The major improved techniques adopted by the townships for conducting the integrated program include:

1. Adoption of newly developed leading rice varieties that are resistant to most of the prevalent diseases and lodging, and are responsive to added fertilization and dense spacing. No more than three varieties are used at each township.

2. Raising of healthy seedlings on improved seedbeds, by means of thin sowing, with optimum fertilization, and careful disease and insect pest control. The seedling nursery is managed cooperatively, with no more than five public seedling nurseries for each unit.

3. Transplanting at optimum spacing, with a recommended spacing of 27 cm. by 13.5 cm. at Meinung and 24 cm. by 15 cm. at the other townships.

4. The rates of fertilizer application recommended for the participating townships are based on the soil fertility of the respective areas.

5. The plant protection practices recommended in a handbook are carried out by the farmers on a cooperative basis.

A study of Table 13-6 shows that the average per hectare yield on the integrated plot was 1,002 kg., or 24 per cent more than that on the check plot. Calculated on the basis of the official rate of NT$4,500 per metric ton, the additional income averaged NT$3,863 per hectare per crop, or 42 per cent for the participating farm families.

The integrated package program, under which the farmers are organized to take part in a joint farming operation to effect an exchange of labor in time of need, makes effective and efficient use of power tillers, sprayers, and other farm implements so that the production

Table 13-6. Rice yield and net income under integrated techniques of cultivation in seven townships

Crop season	Total acreage (ha.)	Average yield of paddy (kg./ha.)			Average net income (NT$/ha.)		
		Integrated plot	Check	Increase in %	Integrated plot	Check	Increase in %
1967-II	2,907	5,000	4,023	24	12,171	8,612	41
1968-I	4,357	5,467	4,411	24	14,309	10,182	41
1968-II	7,724	5,277	4,305	23	12,490	8,587	45
Averages	4,664	5,248	4,246	24	12,990	9,127	42

Source: Plant Industry Division, JCRR.

cost is lowered. This is particularly important since, as rural people are drawn into cities to swell the ranks of the industrial workers, there are fewer and fewer people in the villages to cultivate the land and to provide enough food and fiber to feed and clothe the mounting population.

In addition, these operations provide opportunities to develop local leadership through which improvements in farming, living conditions, and community development can be carried out.

Since joint farming operations have proved so successful in the cultivation of rice, this method has been extended to the cultivation of corn, soybeans, peanuts, sweet potatoes, and oranges, and to radish seed production.

The combining of joint farming operations with integrated improved cultural practices, as successfully developed in Taiwan, may be one of the effective ways to increase crop production through the economy of scale without doing violence to the traditional family farm, which is the cornerstone of rural society in Taiwan.

14. Animal Industry

The first animal projects of JCRR in Taiwan were the hog-cholera vaccine production project of 1949 and the rinderpest emergency control program of 1949–1950. In the latter year an island-wide hog-cholera control program was started. The successful control of animal diseases illustrates the close technical and administrative cooperation between the Provincial Government and JCRR.

The Hog-Cholera Control Program

This program was systematically carried out from laboratory testing to pilot control and to island-wide extension. The responsibility for the program was gradually shifted and finally turned over to the Provincial Government in 1965 after almost 15 years of JCRR assistance.

Hog-raising is one of the most important rural industries in Taiwan. Swine supply protein food for man, furnish manure for making compost to enrich the soil, and provide a major source of revenue for local governments in the form of the slaughter tax. When Taiwan was retroceded to China in 1945, the hog population was only a little more than 500,000 head, hog cholera was prevalent, and the farmers had no interest in raising pigs.

Encouraged by the successful laboratory production of hog-cholera crystal violet vaccine,[1] and particularly by the re-eradication of rinderpest, the Provincial Government started a vaccination program against

[1] C. T. Lee *et al.,* "Experiments on Crystal Violet Hog Cholera Vaccine Production" (in Chinese with an English summary), *Taiwan Journal of Animal Husbandry and Veterinary Medicine,* Vol. I, No. 1, pp. 1–8, June 1950.

hog cholera in 1950 with a subsidy from JCRR. It was gradually found that the selection of virus-donor pigs and the proper disposal of their carcasses had become too much of a responsibility for the Provincial Veterinary Serum Institute, as the vaccination program using crystal violet vaccine grew bigger. Also owing to the short duration of immunity in vaccinated pigs and the high cost of production of the vaccine, it was strongly felt that a new vaccine, preferably one not produced from the original host animal, should be sought.

In December, 1952, a lapinized hog cholera virus was introduced into Taiwan from the Philippines by JCRR. A series of experiments was carried out with the help of the Provincial Veterinary Serum Institute (later transformed into the Provincial Research Institute for Animal Health) to find out the basic character of the lapinized virus and the possibility of its development into a vaccine. After more than a year's study, a fluid vaccine was developed. Pilot field tests were then initiated, and a demonstration project for hog cholera control by the use of lapinized vaccine was finally carried out in Pingtung county in 1954. Its success aroused tremendous interest for the new vaccine among local veterinarians and on the part of the Government.[2]

Pingtung county, situated in southern Taiwan and isolated by the Central Mountain Range and the Hsiatanshui River from the rest of the island, was an ideal site for testing the new vaccine. The success of the Pingtung demonstration project eventually led to the implementation of an island-wide hog cholera control program. The program was moved from Pingtung county northward to areas south of the Chushui River on the west coast and then to northern and eastern Taiwan. In 1957 it was extended to the whole island. Inter-county hog quarantine and meat inspection at slaughterhouses in addition to mass vaccination were enforced. As a result of the effective control of the disease, farmers soon regained confidence and resumed hog-raising. By the end of 1957, the hog population had reached 3.5 million.

For accurate diagnosis and rapid animal-disease reporting, local veterinary diagnostic centers were set up as early as 1950. The services of these centers were improved in the following years as the hog cholera control program was expanded. Extensive mass vaccina-

[2] Robert C. T. Lee, *Hog Cholera Control Program in Taiwan*, JCRR Animal Industry Series No. 2 (Taipei: JCRR, 1953).

tion also required a large number of field veterinarians in order to keep the vaccination level at more than 85 per cent of the hog population. As many as 500 field veterinarians were once either directly or indirectly engaged in the task.

To pave the way for gradually transferring the program to the government, JCRR took pains to consult fully all parties concerned on the matter. Through countless discussions, the farmers and the local governments were finally convinced that the former should be required to pay the cost of the vaccine. This important development made possible the later transfer of the program to the government. At the same time, JCRR persuaded the government to put the veterinary vaccinators employed under JCRR projects on the government payroll so that the control work would not be interrupted after the Joint Commission discontinued its assistance. By 1965 every county or city government in Taiwan had completed the establishment of its own veterinary diagnostic laboratory and had an adequate budget to operate it. Now each local government also has a regular veterinary team to take care of routine animal disease control activities. A well-organized system of veterinary field service has thus been firmly established.

While JCRR was turning over the routine control program to the government, its technical and financial assistance in biologics manufacturing and veterinary research was continued. Funds were granted in 1956 for the establishment of a lyophilized vaccine laboratory at the Provincial Research Institute for Animal Health mainly for the production of freeze-dried hog cholera vaccine. A research hall, tissue culture laboratories, and specific-pathogen-free pig quarters were also constructed with JCRR assistance at the research institute in 1962 and the following years. Moreover, considerable efforts were devoted to advanced veterinary education and training. In 1952–1953, recommendations were made to strengthen veterinary education at the National Taiwan University and the need for a veterinary hospital was pointed out.[3] The result was the establishment of an independent veterinary department to provide five-year professional training, and of a veterinary hospital at the university.

[3] I. D. Newsom, *Livestock Sanitary Control in Taiwan*, JCRR Animal Industry Series No. 4 (Taipei: JCRR, 1953).

Animal Insurance

In close coordination with the development of the disease control program, an animal health insurance program was initiated in 1954 on a strictly self-supporting basis.

The Integrated Swine Program

The integrated swine program initiated in Pingtung county in 1963 has since been extended to the whole island. The program combines all related activities such as (1) vaccination and health insurance; (2) supply of improved breeds through artificial insemination and natural breeding; (3) provision of balanced ration; (4) management improvement; and (5) joint marketing through farmers' associations. It has proved a great success among farmers and led to increased production of swine and better carcass quality.

In the earlier period all such activities as those just mentioned were carried out as individual projects. Credit needed for swine production was provided by farmers' associations and banks, and hog marketing was also handled by the FAs as part of their marketing services.

The swine-breeding scheme in the earlier period was centered on the crossing of Berkshire boars with native sows. Extension of the superior boars was successful because the F_1 hybrids were definitely better pork producers adaptable to the local conditions. Almost every township farmers' association kept one or more Berkshire boars for breeding purposes. Though the practice of artificial insemination for swine had been introduced, few farmers made use of it as its merits were not yet widely known.

In the meantime the Taiwan Sugar Corporation had succeeded in producing three-way-cross pigs by crossing Yorkshire or Landrace boars with the F_1 sows from Berkshire and native. The carcass quality of the hybrid was excellent, but the extension of the three-way-cross piglets by the Taiwan Sugar Corporation to farmers was done on a very limited scale, because it required better feed and better management, both of which were then lacking.

The integrated swine program in 1963 was designed to include all the activities relating to the development of the swine industry. The extension of the improved breed of three-way-cross piglets was accompanied by management improvement and the supply of mixed feed ration. The mixed feed was given to the farmers on credit, and all the pigs were insured to minimize any risks in raising them. The joint marketing of hogs through the farmers' associations completed the process. Under this program, each farmer raised two F_1 sows and 20 pigs as one unit. As it was too expensive for the farmers to keep their own boars, they had to depend on artificial insemination for producing the three-way-cross piglets, thereby learning the advantages of such a practice. The artificial insemination program for swine, which was started some years ago, thus became firmly established.

During 1964–1968, the farmers' associations received help to set up fifteen small feed mixing plants. At first they produced only concentrate feeds to be mixed by the farmers themselves with sweet potato chips and silage according to a proper ratio to form a complete balanced ration. Later on, a complete ration was also manufactured by these plants to satisfy the demand of some farmers whose interest in hog-raising had extended to pure exotic breeds besides crossed ones. Recently, the farmers' associations started a joint project for establishing a modern feed industry to meet the needs of the fast growing swine industry in Taiwan.

The development of the feed industry for hogs is closely related to the progress of the integrated swine program. In Taiwan, the main feed for hogs is sweet potatoes which are grown in most of the paddy rice fields in winter between the two rice crops, and also in dry land. Thus, hog-raising fits in nicely with the local cropping system. Formerly, soybean cakes were distributed by the government to farmers as a protein supplement for hogs. Though the amount of supply was limited at 24 kg. per feeder pig and 48 kg. per sow, this supplementary feed had had a marked effect on the growth of F_1 pigs of Berkshire and native cross. Nevertheless, the incentive was not strong enough for farmers to buy mixed ration to feed their own pigs in those days. For almost ten years the hog population of the island remained between three and three and a half million head.

With the rapid economic growth and continued population expansion in recent years, the demand for pork has been on the increase. As a result, the former conservative policy of basing hog production mainly upon the availability of local feeds has been gradually replaced by one that calls for the development of a truly modern swine industry fully supported by a well-organized feed industry and feed imports as well as such other related activities as meat packing, processing, and marketing.

Poultry Extension Program

The development of a sound business basis has been the main concern of the poultry program since 1962, as it has of the integrated swine program. Private poultry farms have been active in this respect. The liberalization of feed grain imports by the government has also contributed to the rapid growth of the poultry industry. In the early days of the JCRR operation in Taiwan, efforts were concentrated on the importation of breeding flocks and hatching eggs, and the extension of superior breeds, including the Leghorn, the Plymouth Rock, and the Rhode Island Red, for cross breeding in demonstration areas. Disease control projects, mainly on Newcastle disease and fowl cholera, were carried out on a limited scale but in coordination with the improved poultry program. Pullorum testing was done at big chicken farms and hatcheries.

In recent years, farmers' associations have been assisted in implementing a pilot program for the joint marketing of eggs and broilers. Cold storages for eggs, pilot slaughterhouses for broilers and freezing quarters, improved egg crates, etc. have been built or provided. The supplies from the joint marketing program of farmers' associations have helped stabilize egg prices, especially in the egg market of Taipei City.

The duck industry, a traditional rural industry of Chinese farmers, is well established in Taiwan. The native ducks are good layers and the F_1 birds from the Muscovy and the native are excellent meat producers. In the early years, JCRR introduced into the island some Peking ducks and also some Khaki Campells. However, it was found that selection from local flocks would be more practical. Breeding programs under JCRR projects were initiated. The establishment of a Duck Breeding Center in Ilan county was approved in 1967. White ducks have been given special attention in selection, for white feathers will fetch a higher price from feather mills. Artificial insemination has also been practiced in Taiwan in recent years, particularly for the production of F_1 meat ducks.

Turkey-raising is quite popular in Taiwan, especially in the southern parts of the island. Several exotic breeds were tested and imported in the early years for the improvement of the native stock. The Beltsville, a small white bird, is the most promising of all, as it breeds well with the local turkey.

Pastoral Agriculture

Pastoral agriculture on marginal slope land is another example of the application of technical know-how to agricultural production through research and demonstration.[4] In Taiwan where the area of arable land is limited, there remains on the hillsides in mountainous regions a considerable area of lands that can be developed for the combined farming of crops and livestock or for animal grazing purposes. Under the present conditions the practice of raising replacement heifers on hillsides and milk cows on lowland is a stratification approach to dairy cattle production. Demonstration of diversified farming of crops and grasses, through grazing or soiling or by rotation shows the continual search for a still better farming system.[5] Beginning in 1965, JCRR specialists carried out some of the demonstration pastoral programs on the island.

Grass farming, being new to Chinese farmers, made only slow progress. As grass has no economic value unless it is grazed by animals, attention has been given in recent years to a coordinated approach, considering grass farming as a soil/grass/animal complex. Teamwork and cooperative research are emphasized. After a series of tests, several good pastoral farms were established in 1967 and 1968, mostly at govenment experiment stations. Doubts about pastoral agriculture have been finally removed. A project entitled "combined farming of livestock and crops for slope land farmers at Miaoli" was

[4] T. H. Luh, H. W. Ream, and C. Huang, *Grassland Potential and Current Development on Taiwan,* JCRR Animal Industry Series No. 6 (Taipei: JCRR, 1961).

[5] Huang Chia, *Forage Operation Under Small Farm System in Taiwan* (in Chinese with an English summary), Vol. 1 (Taipei: JCRR, 1966).

started in 1966. It is a practical approach to helping slope land farmers in the Miaoli area where they have already done some farming by the conventional method. Their income was extremely low because of the very small size of their farms and low fertility of the hilly land.

Under the project, grasses have been incorporated into the farming system that formerly comprised fruit trees and dryland crops. The grasses, which also serve the purpose of soil conservation, are cut and fed to cattle confined in barns situated at a higher elevation. Manure mixed in water is applied to trees and crops at a lower level by means of small pumps. Thus, not only are the yields of fruits and crops increased because of the improved soil, but farmers obtain additional income from cattle-raising.

The market demand for grazing animals and their products determines the feasibility of grass farming. Since dairy cattle bring higher returns than other ruminants, JCRR has given top priority to planning development of slope lands where there is adequate water for cattle. In recent years, it has assisted farmers' associations both technically and financially in the establishment of pilot milk plants, including collecting and cooling stations, and milk processing plants. An artificial insemination program for cattle and programs for the control of bovine tuberculosis and brucellosis, which were started more than ten years before, have now been turned over to the government. Recently, the production of veal from male calves and beef from culled cows has also been given serious consideration as an important part of the dairy industry. With the increasing demand for beef and milk products, the slope land grass farming program is expected to be widely extended in the future.

Testing Programs

Testing of meat for processing and export is another pioneering program carried out in cooperation with government agencies concerned. It involves institutional changes and improvement of livestock and carcass marketing and slaughterhouse operations. In 1966 the Taiwan Provincial Livestock Research Institute was helped to establish a meat laboratory and assisted the Kaohsiung City Government in the operation of a modern abattoir. Continuous efforts in the past eighteen years have paved the way for the export of pork which may develop into an important business in the future.

As the swine industry has advanced to a more sophisticated stage and the livestock program has been extended from pigs to poultry and cattle, the need for more and better qualified technical personnel and more research in the new fields presents a big challenge.

In the past eighteen years, the Animal Industry Division has placed equal emphasis on research and extension and has actively operated many projects of a pioneering nature. In the early stages of development, the individual project approach was practical because, by that approach, specific problems were solved. The interaction of the various individual projects laid the groundwork for integrated programs. The approval of any individual project has been based on the felt needs of the farmers or on the expectation of providing farmers with incentive to increase production. Demonstration, education, and follow-up field inspection have been emphasized in addition to the imparting of technical know-how. Though integrated programs are much more complicated, they produce greater benefits. The further development of animal industry will require the increased coordination, research, and teamwork which the integrated programs make possible.

15. Forestry

In the first eight years, JCRR efforts were concentrated on (1) the reforestation of depleted windbreak forest lands for the protection of crop production; (2) the reforestation of depleted public as well as private hardwood forest lands on lower elevations for fuel wood and timber production; and (3) an aerial survey of forest resources and land use and the formulation of a sound forest policy and management principles.

The preliminary investigation in 1951 of the forest conditions in Taiwan by Paul Zehngraff, a forest management expert of the U.S. Forest Service and a JCRR forestry consultant, revealed the almost complete destruction of coastal windbreaks and heavy overcut in the huge areas of accessible hardwood forests. Therefore, an expanded reforestation program was drafted by the Division and incorporated into the First Four-Year Economic Development Plan (1953–1956). With the funds provided under the plan, the existing tree nurseries were enlarged, and new ones totaling 200 hectares established. The area of annual planting increased from 10,000 hectares (1945–1952 average) to 33,000 hectares (1953–1964 average) under three successive four-year plans. Several exotic species were introduced, such as slash pine from the United States, eucalyptus from Australia, and Luchu pine from Okinawa.

Kang Han, JCRR silviculture specialist, contributed to the development of the plastic tube planting technique, which is now extensively practiced by government forestry agencies for reforestation in high mountain areas and for extension planting in lower elevation areas. This technique involves little extra cost, but the survival rate of seedlings planted is more than 92 per cent.

To meet the needs of private individuals for capital in making long-term reforestation investments, JCRR since 1964 has had a reforestation loan program. Under it, seven loan projects have been established; loans totaling NT$33,000,000 have been extended to 1,560 tree farmers and one pulp and paper company for planting 1,742 hectares and tending 4,927 hectares. The term of the loans is 15 years, and they carry an interest rate of 8.28 per cent per annum as compared with the prevailing bank rate of 14.04 per cent per annum.

For the purpose of improving forest land productivity and tree quality so as to obtain maximum returns from reforestation investment, a forest genetics seminar was held in 1964 and another in 1965. A 1.5-hectare seed orchard of *Taiwania cryptomerioides* Hay was established in 1967 as part of an overall forest genetics and tree improvement program which has been implemented by a joint committee organized by the forestry agencies concerned.

Forest Resources Survey

Planning and development call for comprehensive and precise information on forest resources. In view of this, an island-wide aerial survey of forest resources and land-use conditions was organized in 1954 with technical assistance of five experts from the U.S. Forest Service. Aerial photographs on the scale of 1:40,000 were employed and 24 strips of 1:10,000 scale aerial photographs were used for sampling purposes. The survey team of 35 Chinese foresters and soil conservation specialists worked on the project for two years. The survey culminated in the publication of 103 forest-type maps on a 1:50,000 scale, and two reports entitled *Forest Resources of Taiwan* and *Land Use Conditions in Taiwan*. The team was later made a permanent unit of the Taiwan Provincial Department of Agriculture and Forestry.

According to the survey, 55 per cent of the total land area or 1,970,000 hectares is covered by forests. Coniferous trees account for 395,000 hectares (20 per cent), broad-leaved trees for 1,461,000 hectares (75 per cent), and bamboo stands for 114,000 hectares (5 per cent). However, most of the valuable coniferous species, which are chiefly virgin and over-matured cypress-type forests located at high altitudes, will be inaccessible until more logging roads are constructed. The bulk of the broad-leaved forests, especially of the tropical hardwood type, are understocked and of inferior miscellaneous species. Therefore, they will contribute least, in terms of per hectare yield, to meeting Taiwan's needs for timber.

Forest Policy

After obtaining the data from the aerial survey, the Provincial Government organized a Forestry Policy Committee consisting of Chinese and American experts. The forest policy and forestry improvement plan as recommended by the committee was officially promulgated in March, 1958, after a formal review and approval by the Taiwan Provincial Government and the Executive Yuan. As this was the first forest policy ever published by the government, it marked an important milestone in the history of China's forestry development.

The policy, as publicly announced, emphasized the balanced development of forest protection and production functions, and urged the government to provide incentives for forest industry and trade development in the private industrial sector.

Forest Road Construction

During 1963–1966, four hundred kilometers of forest and rural roads in five counties were constructed, a project that required in all 4,888,000 man-days of labor. This construction was feasible because of commodities provided under U.S. Public Law 480 (the Food for Work Program); these included 7,887 metric tons of wheat flour and 498 metric tons of edible oil distributed as wages in kind to volunteer workers from the roadside villages. Tools and construction materials were supplied by the people and local governments in the benefited areas.

The roads not only facilitate the transportation of mountain products such as timber, bamboo, bamboo shoots, vegetables, fruits, etc., but also are beneficial to farming, mining, education, communication, and health in the project areas. The mountain products, which used to be carried on human backs to the market, now can be transported by trucks at about one-fifth to one-third of the original cost. The people in the mountain areas, formerly isolated and helpless, have organized committees to collect fees for road maintenance and are learning to better their livelihood through cooperative efforts.

For example, in the Chushan (literally, Bamboo Mountain) area, a 53-kilometer forest road was completed in March, 1967, under this program. This project was sponsored by the Nantou County Government and jointly implemented by three local Bamboo Production Cooperatives. It took two years to complete the construction work by manual labor, with the total U.S. PL480 food commodities distributed valued at NT$8,300,000. The unit cost was therefore only NT$156,000 (US$3,900 equivalent) per kilometer. Now some NT$34,000,000 worth of goods come down annually from the mountains by the truck road, which can save NT$3,500,000 in transportation cost. The bamboo forests made accessible in the area total 6,800 hectares, and 4,800 inhabitants are thus benefited. The many distinguished visitors, Chinese and foreign, who went over to see the project, were all deeply impressed by the enthusiasm and spirit of self-help shown by the rural people in carrying the work through.

Forest Products Research

The forest products research program was started in 1962 in cooperation with universities and research agencies. New methods and new products were developed to meet the needs of the local wood industry. Experimental results already released for commercial production include modified wood products, parquet wood flooring, glued-up turning products, bamboo flooring and paneling, and treated bamboo poles.

Forest Products Market Promotion

Plywood is an important product of the wood industry in Taiwan and its market promotion is effectively handled by the manufacturers themselves. But sawmills and furniture and woodworking manufacturers are mostly small and medium-sized factories without any marketing facilities. To promote the sale of their products, a market survey team was organized and sent to Japan in 1963. Based on its recommendations, a measure for the steady supply of cypress logs for export was worked out and adopted by the government, which has resulted in increased export volumes of locally produced timber in the last five years. The Taiwan Wood and Bamboo Market Development Center, a permanent organization, was established in June, 1964, for the display of current wood products of major manufacturers. It publishes the monthly *Timber-Bamboo Market Bulletin,* conducts wood consumption surveys, and assists foreign and local buyers in finding suppliers to meet their demands.

16. Fisheries

The early fisheries program was a modest one. It included only fish farming and coastal fisheries. In other words, assistance was given only to those fishermen who were part of the rural population. The Joint Commission supported projects for (1) the introduction for culture of exotic and improved varieties of fish and animals such as the Yamato carp and bullfrog, for (2) the installation and improvement of shore facilities such as net treating centers, drying ground, and refrigeration plants for coastal fishermen, and for (3) the training of fishermen.

In 1952 the construction of fishing harbors was added to the assistance program. In extending such assistance priority was given with due consideration to the needs of the fishermen, the benefits to be derived, and arrangements for keeping the harbors in good repair.

In 1953 a project was initiated to train the fishermen in the use of the diesel engines so that gradually the semi-diesel engines could be replaced with the more efficient full diesels. This was the year when JCRR first gave technical and financial assistance to research in fisheries by helping the Taiwan Fisheries Research Institute carry out an experiment on the application of chemical fertilizers in milk-fish ponds. It was also the first time that a loan was made for the construction of a fish cold storage plant.

From 1954 to 1959, the fisheries program was further expanded. The major activities in this period were (1) installation of shore facilities on a province-wide scale; (2) mechanization of small fishing craft; (3) demonstration and extension of the culture of Tilapia in rice paddies; (4) improvement of harbor facilities; and (5) aid to the outlying islands of Quemoy and Matsu.

Expanded activities were carried out in 1957 and 1959 along all lines, particularly in fish culture, mechanization of small fishing craft, and assistance to Quemoy and Matsu. New activities included improvement of fishing gear and methods, installation of refrigeration facilities in fish markets, and construction of small fish salting and drying plants.

With the phasing out of the fisheries program by the ICA China Mission at the end of 1958, deep-sea and industrial fisheries were included in the scope of JCRR aid. To assist the entire fishing industry of Taiwan, a Fisheries Division was created in April, 1959. The technical staff of the new division consisted of a division head, a fishery economist, a biologist, a fish processing technologist, and two fishing technologists. The activities of the Division kept on expanding in the ensuing years, and emphasis was gradually shifted to the development of deep-sea fisheries, including tuna long-lining and trawling, and strengthening of research in marine fisheries and fish culture. A project has been carried out with about a NT$60 million loan a year for the construction of fishing vessels, the procurement of marine engines, synthetic fiber nets, and fish finders, and for the improvement of fish markets and fish ponds.

The following agencies receive assistance in the fisheries program: (1) the Taiwan Fisheries Bureau, which is the provincial government agency responsible for the administration and extension of fisheries; (2) the Taiwan Fisheries Research Institute, which is the provincial government agency for research, experimentation, and demonstration in fisheries; (3) the County and City Governments, each of which has a fisheries section in charge of fisheries administration and extension in the area under its jurisdiction; and (4) the fishermen's associations, which serve the local fishermen and assist the government in administrative and extension work.

Fish Culture Improvement

Because fish culture is very similar to agriculture and is sometimes known as aquiculture, it received early attention. First, the problem of supplying the necessary fish seeds to stock the fresh-water ponds had to be solved. JCRR helped set up a number of carp hatcheries which distributed some five million carp fingerlings per year to fish farmers in different parts of the Island.

Then assistance was given to production of Tilapia fingerlings and demonstration of their culture in paddy fields. The project was at first quite successful; thousands of tons of Tilapia were produced to supply animal protein food to that part of the rural population living in isolated areas where sea fish were not available. With the widespread use of pesticides by rice farmers in later years, however, many of the fish in the paddies were killed, and the rice-cum-fish method of fish culture was practically discontinued.

Milkfish is the most important pond fish in Taiwan. With a total pond area of about 15,000 hectares, almost 30,000 metric tons of fish are produced every year. To further boost the unit area production, the Taiwan Fisheries Research Institute was helped to carry out experiments on the use of chemical fertilizer in milkfish ponds so as to produce more benthic algae, the natural food of the milkfish. Significant success was achieved in another experiment to control the Chironomid larvae which compete with the milkfish for food in the ponds. Findings from these experiments are responsible for increasing the milkfish production in Taiwan, which leads other countries in the technique of milkfish culture.

The Taiwan Fisheries Research Institute was then assisted with a more sophisticated problem—the artificial multiplication of the Chinese major carp, which are the mainstay of the time-tested carp poly-culture in China. After three years of experiment, the research workers succeeded in inducing the carp to spawn, hatching the eggs, and raising the hatchlings to fingerlings. The Republic of China thus became the first country to artificially produce fingerlings of this fish on a commercial scale, and what was formerly imported has become an export item.

Experiment in the application of chemical fertilizers in fresh-water ponds was carried out by the Taiwan Fisheries Research Institute and some county governments under the direction of JCRR's fish culturist. It was found that by the application of superphosphate, the yield of silver carp (a Chinese major carp that feeds on phytoplankton) could be increased threefold. The next step was to help carry out a province-wide extension project. With the assistance of workers paid by the Rockefeller Foundation, research is being conducted on the role of micro-nutrients in fish ponds.

Impressed with the work already accomplished, the Rockefeller Foundation made a grant of US$150,000 to strengthen the fish culture research in Taiwan. This grant covered a two-year period from July, 1966, to June, 1968, and was administered by JCRR. Specialists of the Fisheries Division studied the problems and supervised the implementation of the projects supported by this grant. The research personnel consisted of one limnologist, two chemists, one fish parasitologist, one algologist, and seven research assistants. The fund also financed the services of a Japanese oyster biologist for a period of three months and those of a Japanese shrimp biologist for six months.

Improvement of Harbor and Shore Facilities

Under this early project that is now being phased out, some 30 harbors or anchorages have been constructed or improved with JCRR assistance; they range from large facilities costing NT$10 to NT$60 million to small boat anchorages costing only several thousand New Taiwan dollars each. They have led to the construction of more fishing vessels and thus to increases in the fish catch, but a few of the harbors and anchorages have become of little use in later years as a result of changes in coastal topography.

The public shore facilities JCRR has helped construct consist of concrete-surfaced grounds for the drying of fish and shrimp, net treating centers for the preservation of nets and twines, warehouses for the storage of fishing gear and supplies, signal posts with lights to guide the vessels coming into ports, fueling stations for powered fishing boats, water supply stations to fill the tanks of the fishing vessels, and processing and refrigeration plants. These public shore facilities are all operated by the local fishermen's associations as a service to their members.

Mechanization of Small Fishing Craft

In most developing countries the installation of engines on boats often revolutionizes the business of fish catching. Although all the larger fish boats of Taiwan were powered with either diesel or semi-diesel engines when the Island was retroceded to the Republic of China, as late as 1955 there were still about 7,500 sampans and 14,000 bamboo rafts that depended on human power for propulsion. Owing to the low speed of these small fishing craft, much fishing time was lost, many were lost in rough sea, and their fish catches were small.

In 1954 a project was initiated to extend loans to the fishermen of Ilan County to install engines on 36 sampans. In 1955 a larger loan was extended to install diesel engines on 179 sampans and 20 bamboo rafts. It was reported that the powered sampans caught 2.5 times as much fish as the sampans without power. As a result of this effort, the number of powered sampans in 1965 had exceeded 2,300 or about one third of the total.

The mechanization of the bamboo rafts was at first less successful. The 3-hp diesel engines used were bulky and inconvenient, and likely to break down in rough sea. In 1965 JCRR loaned a few outboard motors made in the United States to selected fishermen in the Ilan area to demonstrate their use. The raft fishermen found that the larger fish catch resulting from the use of the outboards more than compensated for the additional fuel cost. Follow-up action involved the extending of loans to the fishermen for purchasing 238 outboard motors in 1966 and 1967. Funds were also provided as subsidies to impoverished fishermen to buy 80 outboards. At least 1,100 outboard motors were being used by the raft fishermen at the end of 1967.

Mechanization of small fishing craft increased fish production in general, increased the income of the relatively poor inshore fishermen (who comprise the majority of the fishing population) and it saved labor.

Improvement of Fishing Gear and Equipment

Two cases best illustrate how one improvement in fishing gear and equipment led to another. One is the improvement of nets and twines used by the fishermen, another the improvement of fishing lights.

Prior to 1957 all the nets and twines used by the fishermen in Taiwan were made of natural fibers, which are heavy and subject to deterioration as result of chemical and bacterial action. To help the fishermen preserve them, JCRR appropriated grant funds to nearly all the fishermen's associations as subsidies for the construction of net treating plants. These plants were equipped with grinders for pulverizing the Dioscorea roots (the commonly used net preservative) and treating tanks.

With the advent of the use of synthetic fibers in the fishing industry, a program was started to introduce them and extend their use in Taiwan. In 1957 the Joint Commission subsidized the procurement of 27 nylon drift nets. Some fishermen reported that the catch of the nylon nets was two to three times that of the conventional nets. Beginning in 1959 loans were made yearly to help the fishermen purchase synthetic fiber nets and twines, and by the end of 1967 a total of 904,000 pounds of synthetic fibers had been purchased with JCRR financial assistance.

It is estimated that over 80 per cent of the fishing nets and twines in Taiwan are now made of synthetic fibers, which are stronger and lighter in weight than the traditional materials, and resist bacterial decomposition. They do not need preservative treatment, so the net treating plants built with JCRR subsidies have become obsolete and are now generally used for net mending, net storage, and fish auctioning.

The lights used by the fishermen in Taiwan for attracting the phototropic fish to the nets have undergone many changes in the last twenty years. Before 1950 most fishermen used acetylene lamps. Later they used electric lamps powered by electric batteries. JCRR not only made loans to the fishermen for the procurement of batteries, but also provided grant funds to help fishermen's associations set up battery charging stations at seven places from 1952 to 1955. In 1956 the Taiwan Fisheries Bureau was helped to carry out demonstrations on the use of electric generators for fishing lights. The generators were installed on fishing boats and generated sufficient power for lights of 1,000 to 2,000 watts. Because these powerful lamps attracted more fish from greater depths, they were readily adopted by the fishermen. To help the fishermen install the generators, loans were made in 1958 for the procurement of generators. Soon nearly all the fishing boats were using generators, and the batteries and their charging stations became things of the past.

Plastics Replace Bamboo and Wood

The bamboo rafts used by the coastal fishermen of Taiwan are unique. They have distinct advantages; they are cheap, nearly unsinkable, and can be easily beached on the sandy shores. But they also have certain disadvantages. The bamboo-poles that form the raft deteriorate after one or two years of use; since they absorb water, they gain weight and lose buoyancy, and so must be dried from time to time.

To overcome these disadvantages, a project was initiated to help the fishermen experiment on the use of polyvinyl chloride (PVC) pipes to replace the bamboo poles. JCRR is currently supporting a province-wide demonstration of rafts made of synthetic materials, which last three times as long as bamboo poles.

To go one step further, a project has been started to demonstrate the use of fiberglass reinforced plastic (FRP) for making small powered boats and rafts. FRP boats are strong, light in weight and durable (lasting for more than 20 years without maintenance). If this project is successful, it may be as revolutionary a development as the use of nylon nets and ropes.

Improvements in Fisheries Technology

The Taiwan Fisheries Research Institute, with the help of JCRR, has established at Kaohsiung a fisheries technology laboratory which is the only one of its kind on the Island. The laboratory carries out research on canning, salting, and drying of fishery products, fresh preservation of fish, utilization of fish scraps, etc. The following are some of the results:

1. *Manufacture of Fish Soluble*. Research in the utilization of scrap fish and cannery waste for making fish soluble has stimulated local production. While 800 metric tons of fish soluble were imported annually as animal feed before 1962, now twenty small plants in Suao are producing about one thousand metric tons of fish soluble a year, although efforts still must be made to improve the quality of the product.

2. *Fresh Preservation of Fish and Shrimp*. The laboratory has successfully used a phosphate compound to reduce drip loss in frozen

fish and acid sodium sulfite to delay the blackening of fresh shrimp. Both practices are being gradually adopted by the industry and have proved helpful to the fish export trade.

3. *Mixed Feed for Eel.* Eel culture, which has developed into a minor industry with 400 metric tons of fish production a year, is now benefited by a new mixed feed prepared by the laboratory. This mixed feed, with a feed conversion ratio of 1.8, is gradually finding popularity with eel growers. Compared with the trash fish conventionally used to feed the eel, the mixed feed has the advantages of cheapness, convenience of shipping and handling, sanitation, and no need for refrigeration.

Development of Deep-Sea Fisheries

Following the construction of four 350-ton tuna long-liners by the government-operated China Fisheries Corporation and the extension of a loan by the U.S. International Cooperation Administration for the construction of ten 100-ton tuna long-liners, tuna fishing has aroused the interest of the fishermen of Taiwan. In 1961 a loan was used for the construction of twelve tuna long-liners of 160 to 200 tons each to further expand Taiwan's tuna fleet. In 1965 another loan was granted, this time to one single fishing firm, for building a fleet of ten tuna clippers of the 160-ton class. The JCRR loans were extended for the construction of tuna vessels to operate from overseas bases, and practically all the fish they catch are frozen and sold to the United States, Japan, and Italy.

Though there are many skilled fishermen in Penghu (Pescadores Islands), its coastal fish resources are rather limited. To relieve the pressure on coastal fish resources, a loan was made to the Penghu fishermen in 1967 for constructing five 185-ton tuna clippers in order better to utilize the potential human resources there.

The Republic of China has become one of the leading tuna-producing nations of the world. In June, 1967, there were 206 tuna boats ranging from 50 to 1,400 tons each with a total tonnage of 39,609 gross tons. The export value of fish caught by these boats amounted to US$13 million in 1967.

Otter trawling (single-boat trawling) and bull trawling (trawling by a pair of boats) are also considered as deep-sea fisheries. As these boats fish in coastal waters, which are limited in area, JCRR does not encourage the building of more of them. Efforts have been made, however, to increase the efficiency of the trawling gear. A fishing gear specialist has helped the Taiwan Fisheries Research Institute design an otter trawl which catches 30 per cent more fish than the conventional trawl net and is now used by all the otter trawlers in Taiwan.

In view of the successful operation of large Japanese stern trawlers in the distant waters of the Atlantic and Pacific, JCRR financed the sending of nine experienced trawling-boat masters to work on board Japanese stern trawlers in 1966 so that they would gain practical experience and prepare for the day when stern trawling will be started in Taiwan. In 1967 a local fishing firm built a stern trawler of 1,900 tons, which is now fishing off the western coast of Africa.

Fish Marketing and Export

To improve the marketing facilities of Taiwan, JCRR has been extending loans since 1960 for the construction of new fish markets and the expansion and modernization of existing ones, including the installation of refrigeration equipment. There are now 108 fish markets located in various areas of the Island; they sold 374,470 metric tons of fish in 1969.

Table 16-1. Fish production according to categories in the last year of each of the four four-year plans, with 1952 as the base year (in metric tons)

Year	Deep-sea fisheries	Inshore fisheries	Coastal fisheries	Fish culture	Total
1952	18,514	29,696	43,907	29,580	121,697
1956	43,988	63,683	43,259	42,480	193,410
1960	85,210	94,856	30,044	49,030	259,140
1964	126,765	161,151	32,191	56,591	376,398
1968	241,458	208,139	24,891	56,587	531,045

In 1969 fisheries products with a total value of US$44,734,000 were exported, 70 per cent from overseas and 30 per cent from home ports. Frozen tuna ranked first, accounting for 78 per cent, followed by frozen shrimp with 12 per cent.

Fisheries Production in the Last 20 Years

A review of the annual fisheries production of Taiwan in the last sixteen years will reveal the progress made and the general trend of development. Table 16-1 shows that deep-sea fishery production surpassed that of inshore fisheries (fishing with powered boat of less than 20 tons) for the first time by 1968. Coastal fisheries (fishing with nonpowered small craft or without any craft) have gradually lost importance and their production has fallen below that of fish culture, which has remained more or less stable in recent years.

17. Agricultural Marketing and Export

The agricultural recovery in Taiwan went smoothly and the pre-war level of production was reached in 1950–1952—a turning point in JCRR activities and policies. More emphasis has been put on the improvement of marketing since then.

Success in marketing improvement and innovation requires both technological skill and money. Before the existing marketing system and operations can break away from traditional methods that are irrational and inefficient, new attitudes in thinking and new methods in dealing with problems must be available. Financial support to make innovative measures accessible to the parties concerned is equally as important as technical knowledge. The marketing programs, therefore, cover both technical and financial assistance. Since JCRR's role is only to stimulate pioneering and pace-setting innovations, the assistance it provided was mainly to producers' organizations rather than to commercial enterprises.

Modernization of Marketing Facilities

The first and by far the most important aspect of the agricultural marketing program is the modernization and renovation of physical marketing facilities. After the war, the marketing facilities, which had deteriorated because of inadequate maintenance or had become obsolete because of rapid techno-economic progress, were incapable of meeting increased demand brought about by rapid production expansion.

The most popular and important of the various marketing facilities is warehousing for fertilizers and for grains such as rice, wheat, and peanuts. The farmers in Taiwan often deliver their grains for storage to the warehouses constructed and operated by the local farmers' association, of which they are members. The government-controlled chemical fertilizers are also stored in the warehouses of the local farmers' association waiting for distribution to the producers.

Inadequate warehousing facilities inevitably result in physical and quality losses of the stored produce; thus prevention of storage losses by improved warehousing will contribute to the availability of a given output as much as greater production will. With a view to preventing physical and quality losses in storage and to meeting the ever-increasing demand for storage space, JCRR has assisted in the construction of warehouses of better and more efficient architectural design. At present, Taiwan's farmers' associations have about 1,200 rice warehouses constructed mainly of brick and reinforced concrete, with a total storage capacity of about 500,000 metric tons.

Processing facilities such as rice hulling and polishing equipment, peanut husking and corn shelling machines, etc., are very important agricultural marketing facilities in Taiwan. For rice, a two-stage processing, i.e., hulling or husking to produce brown rice from paddy and polishing of brown rice into white rice, has been the standard practice. The local farmers' associations in rice-producing areas are usually equipped with large-scale hulling equipment and rather small-scale polishing equipment. The recovery ratio of rice processing is very important since higher recovery increases the supply available to consumers. Before the war, the recovery ratio of brown rice from paddy was about 76 per cent in terms of weight— from 100 kilograms of paddy, 76 kilograms of brown rice were obtained. Owing partly to the better quality of grain produced, and partly to the more efficient hulling machines popularized through JCRR efforts, the ratio is now about 80 per cent; in other words, 80 kilograms of brown rice can be obtained from 100 kilograms of paddy. Similarly, an automatic silk reeling machine constructed with JCRR support raised the silk recovery from cocoons from 9 per cent to the present rate of more than 10 per cent. Support was also provided to the farmers' associations for the construction of mushroom spawn and feed manufacturing facilities.

Traditionally, small farmers sell their tiny surplus at the farmyard, especially such perishables as vegetables, which are harvested in small quantities almost everyday. Decentralized selling is a source of market imperfection because producers, being remote from the market, tend to be ignorant of market conditions and isolated from market changes. In addition, producers suffer more often than not from inaccurate scales which the buyer brings along and which underweigh the quantity, as well as from payment delinquencies. To improve this situation and protect farmers' interests, assistance was provided for the construction of many fruit and vegetable markets in the production areas to facilitate centralized transactions for both parties. Under this sensible arrangement, the farmers have become more economic-minded producers and sellers, and the dealers more efficient and enterprising merchants.

Since prompt and careful handling is crucial in the marketing of perishable farm produce, collecting and packing houses conveniently situated, properly equipped, and efficiently managed are the starting points of improved marketing operations. Realizing the need for more collecting and packing houses to which farmers could conveniently deliver their produce, JCRR has helped the producers' organizations in the construction, extension, and remodeling of a number of the houses for bananas, oranges, and onions in the production areas. In addition, waxing machines and packaging facilities were installed in some of the packing houses. This equipment is essential to insure prompt treatment and shipping of the mounting deliveries made by producers.

An important portion of harvested grains is retained by the farmers as seeds for the next crop. The seed storage period is usually about eight or nine months for wheat, rape, soybeans, and some types of rice, and during this time the seeds must be kept in good condition with adequate care. With a view to achieving these objectives, assistance has been provided to local farmers' associations in the construction of new-type seed granaries for rice, wheat, rape, soybeans, and peanuts, to which seed farmers can bring their harvest for centralized storage. The cleanness, pureness, and viability of the seeds supplied by these seed granaries have contributed to the increased production of the crops in question.

Accurate and timely market information is the basis for effective decision-making about farm production and marketing; rapid dissemination of such information facilitates the flow of agricultural commodities among different geographical areas and facilitates adjustment of production plans in response to market changes. The net result is an increase in the market value of a given output and a resource employment representing maximum micro-economic welfare. Assistance in this area is represented by the telephone line reconstruction in the business area of the Taichung Fruit Marketing Cooperative, the compilation and publication of market reports, and the transmission of price and arrival information by mass communication media.

Streamlining of Marketing System

The second important aspect of the agricultural marketing program is the streamlining of the marketing system. It calls for the rearrangement of the marketing channels to bring farm produce to the consumers by the shortest possible route; the elimination of superfluous functions to avoid duplication and waste; the combination of functions to enable the marketing enterprises to operate on an economically viable basis; and the readjustment of the sequence in which marketing services are performed so as to minimize incidental costs.

By establishing direct contact between producers and consumers and providing services at cost, cooperative marketing helps eliminate excess profits and makes marketing highly competitive. For this reason, the program has been centered around the promotion of cooperative marketing.

In Taiwan, the most important farmers' organizations engaged in the cooperative marketing of farm produce are fruit marketing cooperatives and farmers' associations. There are six separate marketing cooperatives, with one federation at the top. They control banana and orange exports, from collecting and packing stations down to the port of departure. Previously, practically all banana exports were handled by private exporters, a practice which depressed farmers' prices so much that they did not care to produce more. Presently, the banana export plan requires that 50 per cent of the bananas must be exported by producers' organizations, which are responsible for their physical handling up to the port of shipping.

The farmers' associations have a much wider scope of interest in the cooperative marketing of farm products. By far the most important marketing activities of farmers' associations are related to the collection, storage, processing, transportation, and selling of grains. In recent years, however, cooperative marketing of hogs, poultry, eggs, vegetables, and fruits has been growing in importance.

Hogs, the most important meat source for consumers in Taiwan, are marketed domestically through two channels—the farmers' associations and the profit-seeking commercial establishments. The butchers who buy either directly from the farmers when they operate business in the rural areas or buy at the livestock market when they are established in big townships or cities, slaughter the hogs and retail the pork. Live animals are bought and sold on the market. In view of the need to improve hygienic conditions and expedite slaughter in the abattoir, a system of mechanized slaughter was introduced at Kaohsiung. This is revolutionary in hog-slaughter, but it does not solve the existing problems completely. The sale of hogs as live animals is unsatisfactory, because neither the seller nor the buyer can be absolutely sure of the quality of the hogs on a reliable basis. A new and better system under which carcass transaction replaces live hog transaction is being introduced.

To obviate violent fluctuations in crop production and insure a stable supply of farm products, especially for processing, a system of contract farming has been introduced, with considerable success, for mushrooms and jute. Under this system, farmers contract with canneries for the production and supply of the raw materials at prices agreed upon in advance of the planting season. Farmers' associations establish collecting stations, and centralized collection is undertaken under the supervision of the association personnel to assure fair deals between farmers and canneries.

Rice-Fertilizer Barter System

Rice-fertilizer barter is a unique system in Taiwan's rice marketing. As rice is the major staple food of the people, the government has adopted measures for its control. To procure sufficient rice for the armed forces, government employees, schoolteachers and their dependents, and for sale on the open market (to stabilize the rice price), the government through the Provincial Food Bureau collects paddy approximately equal to 700,000 metric tons of brown rice each year.

This quantity is procured in various ways. Some is a compulsory purchase from paddy-land owners; some is given as payment of the land tax, or as rent or purchase price of public land; part is repayment of rice production loans; and by far the greatest part (about 70 per cent) is barter for chemical fertilizer.

Under the rice-fertilizer barter system, rice farmers receive chemical fertilizers in exchange for paddy at a predetermined ratio. Forty per cent of the barter price is paid on the spot in kind and 60 per cent is regarded as a loan to be repaid after harvest. This barter system benefits rice farmers by assuring them of a reliable supply of fertilizers at stable prices and of a stable and certain paddy price after harvest. It has undoubtedly encouraged rice production since its introduction in 1948. The barter ratio has, however, been gradually adjusted in favor of the farmers. In 1948, for instance, one kg. of ammonium sulphate exchanged for 1.5 kg. of paddy, but in the first crop of 1970 the ratio was 1:0.68, a reduction of 55 per cent. Whereas it took 2 kg. of paddy to exchange for 1 kg. of urea in 1955, it required only 1.09 kg. of paddy to exchange for the same amount of urea of 1970. The fertilizers to be used for crops other than rice are purchased in cash at prices calculated on the basis of the rice-fertilizer barter ratios. The cash price of urea was NT$7,400 per metric ton in 1961 and was reduced to NT$4,650 in the first rice crop of 1970. As only a very small amount of ammonium sulfate was used for crops other than rice, the price of ammonium sulfate was reduced from NT$3,700 to NT$2,900 in the same period. JCRR has contributed significantly to this reduction of the barter ratios and the cash prices.

But as the price of fetrtilizers, both imported and locally produced, has been steadily declining and that of rice rising since 1960, the adjustments made in the rice-fertilizer barter ratio in 1970 has failed to close the widening gap. Consequently, the farmers are still paying too much for fertilizers. To encourage the production of rice and other crops, the government is trying to improve the efficiency of fertilizer production and marketing procedures and to further reduce the prices of fertilizers paid by farmers.

Training Marketing Personnel

In view of the need for upgrading the capabilities of the marketing personnel, educational and training programs are provided to those working with marketing agencies at all levels. There are about 2,650 persons working on farmers' association marketing and purchasing programs of one type or another. They constitute about 31 per cent of the full-time employees of the farmers' associations. Some of them are working on general aspects of marketing, while many are technicians specializing in processing, warehousing, etc. Though the training programs for such personnel vary in duration, subject matter, and emphasis, they generally include two major aspects: general marketing information and specialized marketing techniques. They are usually designed and expected to have pump-priming effects, since it is difficult to provide training programs for all related personnel.

The employees of the fruit, vegetable, and livestock markets are for the most part experienced in operating such markets. They are, however, usually not so knowledgeable with regard to developments in the rapidly progressing technological and managerial sciences. To raise the level of efficiency of the markets that handle large volumes of business, training programs are provided for the key personnel. In the programs, general information about marketing, technical skills of physical handling, and instruction in keeping records are equally emphasized.

Under the existing system, agricultural products for export are inspected by the Commodity Inspection Bureau, first at the local shipping point and second at the port of export. But since the inspection is done on a sampling basis, some low quality goods might well be sold on the international market. To correct this situation, training programs are provided for the quality-control personnel who are employed in the packing houses for perishables established and operated by the growers' organizations. The main objective is to acquaint the trainees with detailed knowledge of the standards and grades established by the Central Government. The technical personnel of the Commodity Inspection Bureau are also trained to bring their knowledge and skills up to date. The combined effect of these programs is to eliminate the loopholes through which goods of inferior quality might flow into foreign markets.

To protect the public interest, all marketing agencies and organizations must operate according to the provisions of relevant laws and administrative orders. To see to it that these regulations are properly observed, the Provincial Government and the county governments are empowered to supervise the marketing agencies and market organizations operation within their jurisdiction. JCRR has helped the various government agencies concerned to provide supervisory, consultant, guidance and teaching services.

Introduction of New Scientific Information through TA Training

JCRR has sent marketing personnel abroad for advanced studies and practical training. The participants were sent mostly under the Sino-American Technical Assistance program and the Sino-Japanese Technical Cooperation program. With a training period ranging from about three months up to one year, the programs emphasized both theoretical aspects and practical problems. The TA training was at first mostly carried out in the United States and Japan, but recently some trainees have also been sent to the European countries. Holding key positions in government agencies or private organizations, the returned trainees disseminate the newly acquired knowledge and techniques in handling practical business matters and planning marketing programs.

The dispatch of special missions to foreign countries for study and observation was another means by which new knowledge was acquired. Though small in size and relatively short in duration, the study missions gained information in solving practical and immediate problems and the members on their return could demonstrate the new ways of handling things that they had learned. Participation in international conferences and workshops had similar significance.

JCRR invited R. F. Wilcox, former general manager of Sunkist Growers Inc. and present director of the Central Bank for Cooperatives and chairman of the National Association of Farm Cooperatives in the United States, to advise on fruit and vegetable marketing, particularly for export. Wilcox and other experts, being free from any preconceived ideas, are able to see things clearly, and have been able to recommend practicable improvement measures within a short period of service.

Foreign visitors who come primarily to investigate the conditions here or to participate in various kinds of training programs have also contributed information that, though necessarily piecemeal and incomplete, provides a useful supplement to already available data.

Marketing Surveys and Market Research

Recognizing the importance of marketing studies, JCRR has made investments, both financial and human, in survey and research programs on marketing. The programs were carried out by the National Taiwan University, the Provincial Chunghsing University, the Provincial Department of Agriculture and Forestry, the Taiwan Agricultural Foreign Market Research Center, the Sino-German Socio-Economic Research Institute in Bonn, West Germany, and by JCRR specialists.

Before the war, marketing studies were practically monopolized by academicians, their subject and content dominated by academic curiosity. After the war, the scope of the studies was much expanded; they became the concern of not only academic circles but also of government agencies and private organizations.

Initially, marketing studies were planned to shed light on the existing marketing situation. They were centered around the identification and description of marketing channels, the nature, kinds, and functions of marketing agencies and market organizations, physical facilities, marketing practices such as collecting, storage, processing, transaction, transportation and other physical handling of the commodity, and marketing costs and margins. These descriptive studies were useful for understanding the overall picture of the marketing economy and providing a basis for further studies in detail. Surveys and research of this nature conducted in the past concerned rice, wheat, peanuts, hogs, bananas, vegetables, pineapples, tea, sweet potatoes, poultry, and eggs. Many of them covered only marketing proper, but some of them were overall economic studies dealing with production, marketing, consumption, and international trade.

Then the major interests of marketing research shifted from general descriptive studies to the analysis of problems which hindered progress in marketing. These studies were aimed at raising marketing efficiency, rationalizing marketing practices, and modernizing marketing operations. Though relatively narrow in scope, they contributed to the solution of immediate problems.

In recent years, keen interest has been shown in market studies, especially in the study of foreign markets. As the promotion of agricultural exports is considered urgently necessary for the further development of the national economy, several market studies for Taiwan's exportable commodities have been made. They have dealt mainly with market conditions in Japan and the countries of the European economic community.

Promotion of Export Trade

As an island with limited natural resources and a small domestic market, the economy of Taiwan has to depend heavily on export.

The value of all agricultural exports from Taiwan was about US$114 million in 1952 and it had grown to US$292 million (excluding US$44 million from the export of plywood in 1967). With the foreign exchange thus made available, it has been possible to import machinery and raw materials for industrial development. Before 1963 there was always an unfavorable balance of trade. Without the foreign exchange brought in by agricultural exports, the deficits would have been greater. Therefore, Taiwan's agriculture has not only contributed to its general economic development, but also improved its international payments position.

The export of agricultural products, especially processed products, has also been promoted. Before 1957, can bursting, caused mainly by improper seaming of cans, was a serious problem adversely affecting the export of canned food items. To overcome this difficulty, a number of can seaming projectors were obtained from the United States for use by the Provincial Bureau of Commodity Inspection and Quarantine in checking the products of local canneries. This and other forms of assistance, including technical training, have enabled the food processing industry to improve and grow rapidly. In 1960 the foreign exchange earnings from the export of canned foods totaled only US$8.5 million. By 1967 the amount had increased to US$78 million, which was 12 per cent of the total foreign exchange receipts in that year.

Canned pineapple is one of the main export items of Taiwan. Some trial sales of the product to the United States were made before 1959, but the shipments were often rejected owing to the high mold count in the canned crushed pineapple. At the request of the Chinese Government, JCRR organized a technical committee to supervise mold control and environmental sanitation improvement in factories. As a result, only a year later Taiwan canned pineapples were successfully sold on the United States market.

Production of canned mushrooms was started on a trial basis in 1959. Improper processing techniques led to discoloration of the mushrooms, the blackening of the inside of the cans, turbidity of fluid, and can bursting. Export of this item was therefore out of the question. A program was launched quickly for improvement by assisting in introducing better processing methods, conducting training courses for farmers in mushroom harvesting and shipping, establishing a mushroom inspection station, promoting planned production, and establishing a sound marketing system for mushrooms. Gradually, the quality of canned mushrooms produced in Taiwan has approached the international standard. In 1967, their export earned US$33 million in foreign exchange.

Asparagus is another important export item. In 1963 only US$4,000 worth of it was exported, and the export value reached US$24 million in 1967. This canned food item was first produced with JCRR assistance in 1961 for domestic consumption only. In 1964, its production was increased for export, and its quality has since been improved.

Worthy of particular mention is the export of bananas. In the five-year period from 1963 to 1967, the amount of foreign exchange earned annually from banana export increased more than seven times, from US$8.6 million to US$62.0 million. Japan has always been the biggest market for Taiwan bananas. In April, 1963, the Japanese government announced the liberalization of banana import. Since then efforts have continually been made to increase banana production and improve the quality of the fruit so as to maintain and further strengthen the position of Taiwan bananas in the Japanese market. In cooperation with the Taiwan Provincial Department of Agriculture and Forestry, JCRR in 1963 drew up a plan for boosting production and promoting quality improvement of bananas. The steps taken under this plan included the application of soil conservation practices on slope lands that could be used to plant bananas so that the banana acreage could be expanded; establishment of demonstration orchards for the extension of improved cultural techniques; effective control of pests and diseases; promotion of the use of bamboo poles as banana support against typhoons; improvement of packing and transportation facilities; and extension of production loans to banana growers.

Before 1963 over 95 per cent of all bananas exported from Taiwan had been handled by fruit exporters, with the middlemen taking the lion's share of the profits. The banana growers could get less than half of the profits of export sales and therefore lost interest in growing the crop. This constituted a serious obstacle to any increase in production and any improvement in the quality of bananas. With the liberalization of Japan's banana import in 1963, the Foreign Exchange and Trade Commission (FETC) of the Chinese Government established a Banana Production and Marketing Guidance Committee in charge of boosting production, increasing the banana growers' share of profits and improving the marketing system. FETC appointed Y. S. Tsiang, commissioner of JCRR, as head of the Committee. The Joint Commission gave its full support to the program, technically and financially.

A new marketing and quotation system was introduced and exports were made on a planned basis. Under the new system farmers did not have to depend on the exporters to sell their bananas to Japan. Producers and banana growers' cooperatives were allowed an export quota of 50 per cent of the amount of bananas exported. The other 50 per cent was allocated to fruit exporters. Consequently, the banana purchase price was raised, and the banana farmers' income increased immediately from an average of NT$102 to NT$203 per basket (100 lbs. per basket) of the fruit. Farmers began to grow more and better quality bananas. As a result, the banana acreage increased from 18 thousand hectares in 1964 to 44 thousand in 1967; and the total production from 268 thousand metric tons to 654 thousand in the same year. The foreign exchange earnings from banana export increased from US$33.3 million in 1964 to US$62.0 million in 1967, as shown in Table 17-1. But despite technological improvements, the yield per hectare remained practically the same from 1964 to 1967. Only in 1965 when there was no typhoon and farmers began to adopt new improved methods increasing the number of banana plants per hectare by close spacing did the yield increase to 16,765 kg. per hectare. Because of two serious typhoons in 1966 and also in 1967, the average yield of bananas dropped to less than 15 thousand kg. per hectare.

Table 17-1. Production and export of Taiwan's bananas

Year	Harvested area (ha.)	Total production (M.T.)	Yield per ha. (kg.)	Total export US$ million
1964	18,086	267,898	14,813	33.3
1965	27,443	460,094	16,765	55.3
1966	36,512	527,721	14,453	52.6
1967	44,107	653,800	14,823	62.0

Sources: Taiwan Agricultural Yearbook, the PDAF, 1968; Export and Import Exchange Settlements for the Year 1967, Bank of Taiwan.

The following facts are also noteworthy:

1. Sugar and rice were formerly the main export items from Taiwan, together accounting for about 80 per cent of the total annual export earnings. In recent years Taiwan's export trade has been considerably diversified. Yet, even though industrial products have been exported in increasing quantities, agricultural commodities have remained an essential component of the export trade.

2. For export of every US$1 worth of agricultural goods, there is a net foreign exchange receipt of about US$0.80 on the average. In the case of industrial goods, the net receipt is only US$0.50 because foreign exchange has to be expended to import raw materials and machinery for their manufacture. Agricultural exports still account for a major portion of the total net foreign exchange earnings.

3. New export crops have appeared from time to time in recent years, and the development of their export has been very rapid.

4. The agricultural products exported from Taiwan are mostly crops grown on slope land and marginal mountain land by inter-cropping and rotational cropping methods. Hence, they do not compete with food crops for land.

5. The export of agricultural products increases the cash income of the farmers who, as a result, are able to save their money and use their savings to improve their farms.

6. The production of export crops absorbs much of the surplus labor in rural areas and gives the farm families additional income.

7. Farmers gain additional income not only from the production of export crops, but also from their by-products. For example, the price of rice straw has gone up because it is in great demand in mushroom culture. Bamboo and straw rope which are used in banana production and packing now also fetch higher prices than before.

8. The export of agricultural products means that they are being increasingly commercialized. This has led farmers to pay more attention to technical improvements and to emphasize quality more than quantity.

9. The expansion of agricultural export trade has promoted the development of farmers' organizations, whose marketing services have shown considerable improvements in recent years.

18. Agricultural Statistics and Economic Studies

Agricultural statistics and economic investigations have become indispensable adjuncts of economic development and planning. Through both independent and sponsored projects, JCRR has carried out many statistical and economic activities.

Agricultural Marketing and Prices

Marketing costs of various important farm products and their price spreads between markets have been studied since 1951 to evaluate the efficiency of the marketing services rendered by middlemen. The information on price spreads supplies a basis for determining the profits retained by the middlemen. Loans have been made to various marketing agencies for strengthening their activities and functions. Because of the rapid expansion of international trade of agricultural products, equal emphasis has been laid on foreign market study in recent years.

Under a trial project, JCRR undertook in 1953 to compile a series of indices of prices received and paid by farmers in Taiwan. The study was made in an effort to arouse government officials to an awareness of the need for publishing data in this field and developing methods for the compilation of a cost-of-living index for farmers. As a result of this trial project, the Provincial Accounting and Statistics Bureau compiled and published *Monthly Statistics on Price Received and Price Paid by Farmer in Taiwan* in cooperation with the Provincial Department of Agriculture and Forestry, the Provincial Food Bureau, and JCRR.

Farm Management Research and Record Keeping Program

Apart from technological and institutional improvements, the efficiency of farm management is also important in determining the success or failure of farm operations. Research and extension of better farm management by public agencies and farmers' associations is no less important than technical improvements in achieving bigger output on the farms. Farm management specialists and extension workers can render a valuable service by helping producers plan farming systems that will bring the highest returns. For example, a survey of the production costs of the major crops grown in Taiwan has been supported for several years. The survey has provided information about the comparative advantages of each crop and cropping system in different regions, thus enabling the farmers to choose crops best suited to their farms.

The farm record keeping program during the past 15 years is also an important step in helping the individual farmers improve their farm planning and operation. Since 1960, the farm record keeping systems have been primarily for collection of economic data and training. Six local farmers' associations chosen from north, central, and south Taiwan have been designated to supervise some 100 farmers in keeping daily records; the Provincial Department of Agriculture and Forestry has been responsible for posting and compiling the data obtained from daily records. Each farmer participating in the program is provided with a farm account book. He keeps a record of his farm operations throughout the year with the assistance of extension workers of the township farmers' association. The local farm extension workers call on each farmer several times in a month and check the completeness and accuracy of their books. At the end of the year, all the account books are checked again, and the results carefully tabulated and analyzed. A report on the tabulation and analysis of farm records has been published each year by the Provincial Department of Agriculture and Forestry. In addition, a balance sheet of each farmer's annual farm income, with commentary on the weaknesses and strengths of the farm operations, is sent to him as a reference for planning his farm organization and operation practices in the next year. The local extension workers may use the information from the program in discussing with farmers their farm management and operational problems. In this way, the farm management specialists and extension workers play an important role in working for better farm management and operations. This program has also provided material for making further studies on farm management as well as for compiling national income statistics.

Crop Reporting and Agricultural Census

Reliable statistics on crop production are essential to any country which has either surplus farm products for export, or import requirements to meet the deficits in raw materials for industries and food processing. Such information is also essential to economic planning, which requires objective and reliable crop data. With these ends in view, JCRR has been assisting the Provincial Department of Agriculture and Forestry to have accurate and timely statistics by improving their crop and livestock reporting system. JCRR assistance was given largely in the form of providing land maps, equipment, and vehicles to the reporters at all levels; a training program was also provided for all government workers involved in agricultural reporting.

In addition to the agricultural reporting system, a cost of production survey has been supported. It is sponsored by the Provincial Department of Agriculture and Forestry; its function is to analyze any structural changes in the farm production cost and to help the Department collect the cost data needed for policy decisions and production efficiency evaluation.

To secure information on changes in agriculture during the past ten years, the Provincial Department of Agriculture and Forestry, the Provincial Food Bureau, and JCRR jointly undertook in March, 1956, the first sample census of agriculture since the restoration of Taiwan to China. By using the method of 5 per cent random sampling, some 4,000 neighborhoods were selected and each of the 37,636 farmers in them was interviewed. The questionnaire included some 100 questions related to the number of families, farm size, tenure, labor, employment, crops, livestock, irrigation, equipment and implements.

Nearly 1,000 persons were recruited as enumerators, team leaders, and supervisors. As a result, *Report on the 1956 Sample Census of Agriculture, Taiwan, Republic of China* was published.

Following the 1956 census, JCRR cooperated with the Provincial Department of Agriculture and Forestry and the Provincial Food Bureau to make another agricultural census in 1960. It marked the first participation of the Republic of China in the program of the World Census of Agriculture and also the first experience of the government in carrying out a complete enumeration of all farm households in Taiwan; the results were published in *General Report on the 1961 Census of Agriculture, Taiwan, Republic of China.*

The program was considered an integral part of the 1960 World Census of Agriculture sponsored by FAO. Besides financing about two thirds of the program expenditures, JCRR provided active technical aid. For collecting statistical information, two main field enumerations—the complete census and a 10 per cent sample census—were undertaken for 1960. In the complete census, some 850,000 identified farm households were visited by about 5,000 enumerators selected from township offices and farmers' associations. A questionnaire covering the three main topics of agricultural population, land and its utilization, and crop area was used in taking this census. Since it was the aim of the 10 per cent sample census to secure more detailed information to supplement the data of the complete census, about 80,000 sample farm households selected at random from the farm households of the complete census were revisited by 1,200 competent and well-trained enumerators. The more detailed questionnaire used in the sampling census contained ten items: (1) agricultural population; (2) land and irrigation facilities; (3) crop area; (4) agricultural labor; (5) livestock and poultry; (6) by-products (forestry and fisheries); (7) fertilizer and soil improvement; (8) farm tools, equipment and installations; (9) chemicals; and (10) others.

Another 5 per cent sample census of agriculture was also supported in 1965 and published as *Report on the 1966 Census of Agriculture, Taiwan, Republic of China.* To obtain serial information on agricultural basic data for use in planning social and economic development and making agricultural policy, it was decided to conduct a complete census every ten years, with a mid-decade sample census. The complete census for 1970 is now under preparation by the Ministry of Economic Affairs.

Farm Income Survey

Just as the national income is a barometer for interpreting economic thought and action, so is farm income (a segment of the national income) a yardstick for evaluating and determining agricultural policies and efforts for the improvement of farm life. In view of the necessity of obtaining some farm income data for measuring Taiwan's farm economy, JCRR initiated an island-wide farm income survey in 1952.[1] The survey covered 100 townships and included 4,000 farm families selected from the 13 agricultural regions on this island. This survey was the largest of its kind ever undertaken in Taiwan. Following methods used in the survey of 1952, JCRR conducted three more in 1957, 1962, and 1967[2] that not only turned out raw statistics of income and expenditures on farms but also provided basic information on changes in the farm economy over the years.

Economic Research

A comprehensive socio-economic study undertaken in 1953 by A. F. Raper and a similar one by E. S. Kirby in 1959 provided detailed information on rural conditions, attitudes, and problems, and also provided a basis for further programs to meet the needs of farmers.[3]

The socio-economic study of 1959 covered 18 townships with 75

[1] Y. C. Tsui and S. C. Hsieh, *Farm Income of Taiwan in 1952*, JCRR Economic Digest Series No. 4 (Taipei: JCRR, 1954).

[2] Y. C. Tsui, *A Summary Report on Farm Income of Taiwan in 1957 in Comparison with 1952*, JCRR Economic Digest Series No. 13 (Taipei: JCRR, 1959). See also Y. C. Tsui and C. Y. Teng, "A Report on Farm Income of Taiwan in 1962 in Comparison with 1957 and 1952," mimeograph (Taipei: JCRR, 1965), and Sung Mien-nan and Yu Yu-hsien, *A Report on Farm Income of Taiwan in 1967* (in Chinese) (Taichung: Provincial Chunghsing University College of Agriculture, 1969).

[3] Arthur F. Raper, *Rural Taiwan—Problem and Promise* (Taipei: JCRR, 1953) and E. Stuart Kirby, *Rural Progress in Taiwan* (Taipei: JCRR, 1960). See also Jean T. Burke, *A Study of Existing Social Conditions in the Eight Townships of the Shihmen Reservoir Area*, JCRR Economic Digest Series No. 14 (Taipei: JCRR, 1962).

households in each township or a total of 1,350 households selected for investigation. In addition, information was collected through local farmers' associations and other related agencies. Although the problems of social structure and adjustment in Taiwan have their own intrinsic importance, they have a wider significance, since Taiwan is linked historically, culturally, sociologically, and economically with the East Asian region as a whole.

JCRR also gave close attention to agricultural economic development. An analytical review of agricultural development on Taiwan, using an input-output and productivity approach, was made in 1958 and the report published in the Economic Digest Series.[4] This study covers the period 1910–1956, with a projection of agricultural development in 1960 and 1970. It shows that the average annual growth rate of agricultural output (crops and livestock) at different stages of agricultural development was as follows:

Average of prewar period	1910–1939	3.43%
Average of war period	1939–1945	−12.33% *
Average of postwar period	1945–1956	9.90%
Average of whole period	1910–1956	2.70%

* Owing to the damage done by Allied bombing to agriculture in the later years of World War II, the average annual growth rate of agricultural output declined to −12.33 per cent during the war period.

As pointed out by this study, two factors affecting the expansion of agricultural output were the increase of input factors and the improvement of technology. Taking 1935–1937 as the base, the aggregate input in agriculture increased by about 16 per cent, while output increased some 36 per cent, or more than double the input increase. This suggests that agricultural technology has also advanced rapidly with commensurate improvement in efficiency and productivity.

Following the analytical study of Taiwan's economic development, a further study on agricultural development and its contributions to

[4] S. C. Hsieh and T. H. Lee, *An Analytical Review of Agricultural Development in Taiwan—An Input-Output and Productivity Approach*, JCRR Economic Digest Series No. 12 (Taipei: JCRR, 1958).

economic growth in Taiwan was carried out and published by JCRR.[5] This new study served as an important source of reference for the Seminar on Agricultural Development held in Taipei in June 1966 under the joint sponsorship of JCRR and the USAID.[6] Fifty-five representatives from seven Asian countries, staff members of the AID missions attended the Seminar.

As Taiwan has steadily achieved high rates of agricultural growth during the postwar years, particularly since 1952, it was thought that its experience might be of some practical value to other Asian countries which are trying to develop their own agriculture. Therefore, the Seminar was held to familiarize the participants with the achievements the Republic of China has made in agriculture in its island province of Taiwan and to consider how and to what extent Free China's accumulated knowledge and past experience might be applicable to other developing countries facing similar, if not identical, agricultural problems.

In 1958 an analysis of the interdependence of agriculture and other sectors of the national economy was initiated.[7] The purpose of this trial compilation was to indicate the input-output relationship between agriculture, industry, and export trade on Taiwan. This study furnished information relating to (1) the flow of products from agriculture to other industrial sectors; (2) the effects of increase or decrease in domestic and foreign demand for farm products; (3) the flow of products and resources to and from agriculture; and (4) the effects of decrease in farm prices on the prices of other goods in the industrial sector. However, since current data were scarce, and budget and personnel limited, this study was carried out on a rather small scale and restricted to nine industries.

In recent years the techniques and methods applied in the field of economic development planning have entered a new stage. The input-

[5] S. C. Hsieh and T. H. Lee, *Agricultural Development and Its Contributions to Economic Growth in Taiwan,* JCRR Economic Digest Series No. 17 (Taipei: JCRR, 1966).
[6] JCRR, *Summary Report of Seminar on Agricultural Development* (Taipei: JCRR, 1967).
[7] S. C. Hsieh, T. H. Lee, and Y. T. Wang, "Interdependence of Agriculture and Other Sectors in Taiwan" (in Chinese), *Industry of Free China,* Vol. 15, Nos. 1 and 2 (Taipei: IFC Publication Committee, 1961).

output table analysis has been used in formulating economic development plans by the Council for International Economic Cooperation and Development (CIECD), which compiled an input-output table of 54 industries for 1961, and a similar one for 1964. Included in the agricultural industries were rice, other common crops, sugar cane, crops for agricultural processing industries, other horticultural crops, hogs, other livestock, forestry, and fisheries. JCRR was requested to compile that part relating to the agricultural sector for both 1961 and 1964.

In the study on the effects of population trends on economic development in Taiwan, JCRR has emphasized the analysis of relationship between population growth and economic development as reflected in the requirements of saving, investment, and manpower. It has conducted studies on population and food problems such as the food administration system and the marketing of rice and other grains, the long run projection of demand for food and main agricultural products, the food consumption survey, and other similar studies on population growth and food problems in order to furnish the government with basic data and offer suggestions to the government for its food program. Based on these studies and surveys, a food balance sheet for Taiwan starting from 1935 has been compiled and published by JCRR.

The population of Taiwan today is characterized by rapid natural increase and youthfulness. The distribution by age groups suggests that people under 14 years of age account for more than 44 per cent of the total population. A calculation of mortality and fertility shows that the population of Taiwan will be doubled in less than 24 years, and its total number will reach 14 million in 1970, 19 million in 1980 without family planning, but only 17 million in 1980 with family planning. The high growth rate of Taiwan's population has been a threat to its economic and social development. Discussion of this problem in the past was nevertheless focused largely on such aspects as family planning and supply and demand for food. This problem has been tackled from the viewpoints of consumption, savings, investment, labor supply, and employment.

Since labor supply is always an essential factor in industrial development, JCRR and the Chinese Land Economics Research Institute made a small sample farm labor survey in the Taichung area in 1957 to obtain a general idea of the nature and scope of human resources use and the volume of surplus labor in rural Taiwan which could be transferred to nonagricultural jobs. In addition to this farm labor survey of 1957, JCRR cooperated with the Provincial Chunghsing University in another survey in 1960. Whereas the 1957 survey was directed toward investigating the potential demand for, and supply of farm labor, the 1960 survey studied the records kept by farm families to find how they had used farm labor. Sixty farm households were selected from six townships in the Taichung area. Each household selected was given a book to record daily farm labor use for a period of eight weeks, some of which were in busy and some in off seasons. However, in order to find out how farm labor was used throughout the whole year and thus determine the labor structure in the production of major crops, 12 farm households selected from the 60 were asked to record family labor use by items and by hours of work daily throughout the whole year. This study provided a detailed picture of labor utilization.

Another study on the effects of population pressure on the crop pattern was also carried out in 1964. The linear programming method was employed to analyze how the intensive cropping system, which is popular in Taiwan, affected Taiwan's rural labor employment. It was found that, although the system did increase employment and income, it seemed incapable of absorbing all the surplus labor in rural areas. In order to trace the movements of that portion of the surplus labor which was not absorbed, JCRR conducted a further study on rural labor mobility in 1964.[8]

[8] Y. C. Tsui and T. L. Lin, *A Study on Rural Labor Mobility in Relation to Industrialization and Urbanization in Taiwan,* JCRR Economic Digest Series No. 16 (Taipei: JCRR, 1964).

19. Agricultural Credit

The establishment of a sound agricultural credit system through institutional improvement has been the major goal in implementing agricultural credit programs in Taiwan. In this process, private lending to farmers was gradually replaced by institutional credit. In 1967, as much as 81 per cent of the farmers' credit needs was financed by the agricultural banks, JCRR, farmers' associations credit departments, and government enterprises, with the other 19 per cent coming from private sources such as merchants, relatives, and friends. This was a substantial change from the situation in the late forties when private money lenders played a predominant role in agricultural credit and financing, supplying 82 per cent of the agricultural credit for farmers in 1949.

The change has come about through the implementation of a series of credit programs beginning with the Experimentation of Farm Operation and Improvement Loan Program of 1955. It was followed by the Farm Credit Demonstration Program in the next year and the Supervised Agricultural Credit Program of 1958. Finally, the Unified Agricultural Credit Program was launched in 1961 and is still continuing.

Since the first three credit programs have been described in a previous work by the writer,[1] only the current program will be discussed and its recent developments reviewed here.

When the Unified Agricultural Credit Program was inaugurated, it was enthusiastically welcomed by the farmers and the farmers' associations to which they belonged. High-ranking officials of the In-

[1] T. H. Shen, *Agricultural Development on Taiwan since World War II* (Ithaca, N. Y.: Cornell University Press, 1964), pp. 310–323.

ternational Cooperation Administration China Mission and the Council for United States Aid made a field trip to several township farmers' associations to interview individual farmer-borrowers and to check the project operation in June, 1961. Upon their return to Taipei, the ICA officials summarized their impressions: "The new program marks the start of a much needed credit organization capable of providing capital to meet particular needs of farmers in Taiwan and will play an important role in the accomplishment of the agricultural development program. In the long run the effect may be favorably comparable to the increase of production which resulted from the Land-to-the-Tiller Program." [2]

William I. Myers, former dean of the College of Agriculture, Cornell University, and former governor of the U.S. Farm Credit Administration, who served as short-term consultant to ICA and JCRR in 1960, and then in 1962, had this to say about the Unified Agricultural Credit Program after his inspection of the project operation in February, 1962: "The Unified Agricultural Credit Program has made an excellent start in its first year of operation and the outlook for 1962 is favorable. The plan of organization and operation follows the outline prepared in 1960 and seems to be generally satisfactory. The establishment of the Agricultural Credit Planning Board is a desirable step since it makes possible amendments in operating rules and regulations justified by practical experience. As always, unforeseen problems have arisen in actual operations but these are relatively minor ones that can be solved satisfactorily by the staff and the Board. The regulations adopted to limit the use of funds of the Credit Divisions [i.e., the credit departments of the farmers' associations] seem reasonable and necessary to protect their financial soundness. If these plans are carried out intelligently, this unified agricultural credit system will be the best in southeastern Asia and will serve as a model for other countries. Most importantly, however, it will make a vital contribution to agricultural development and productivity in Taiwan." [3]

[2] JCRR News Release No. 2469, August 12, 1961.
[3] Official report to the Joint Commission on Rural Reconstruction (1962) (unpublished).

Progress of the Unified Agricultural Credit Program

In all 259 townships farmers' associations, representing 88 per cent of all the township farmers' associations that provide credit facilities to their members, have been enrolled in the Unified Agricultural Credit Program from FY1961 through FY1968. Thus, almost nine out of ten farm families in Taiwan have access to the supervised credit service.

Under the program, the farmers' association is required to mobilize its available funds for the financing of agriculture and to supervise all uses of loans by the farmers. To implement the program, an Agricultural Credit Fund was established, with appropriations amounting to NT$223.4 million from the United States aid counterpart fund in the FY1961–1965 period. The Fund provides each participating association with two kinds of loans: direct and indirect. Direct loans are given without interest charges to the farmers' association to match its lending capital. When the farmers' association has fully utilized all its lending resources and still needs additional funds for making more loans to farmers, it can borrow from the Fund at cost, indirectly, through the Land Bank and the Cooperative Bank.

With the phasing out of United States aid to China at the end of FY1965, the appropriations from the counterpart fund were discontinued but, as originally scheduled, the first batch of the participating farmers' associations began to make their first installment repayments to the Fund. Since then, the capital repayments from the direct loans and the interest payments from the indirect loans have been used as a revolving fund to finance more farmers' associations. The aggregate of loans made from the Agricultural Credit Fund directly and indirectly to the farmers' associations from FY1961 through FY1968 amounts to NT$275 million.

By December 31, 1968, the loans made by the participating farmers' associations to their farmer members totaled NT$5,040 million. With 319,879 borrowers, more than one of every three farm families in Taiwan has benefited from the program. They have used about 39 per cent of the loans for purchase of farm land and land improvement, 37 per cent for purchase of farm tools, machinery, and draft cattle and the construction and repair of farmhouses, and 24 per cent for

crop, livestock, and sideline production. The rate of interest charged
has been gradually lowered, as shown in Table 19-1.

Table 19-1. Rates of interest of unified credit loans to farmers

Date on which interest rate became effective	Monthly interest on unsecured loans (%)	Monthly interest on secured loans (%)
May 1, 1961	1.50	1.35
Aug. 8, 1962	1.50	1.32
July 1, 1963	1.32	1.11
March 1, 1964	1.26	1.11
Feb. 14, 1966	1.20	1.11
May 6, 1967	1.11	1.05

Source: Agricultural Credit Division, JCRR.

Nearly 25 per cent of the loans were made for a term of up to 12
months and 75 per cent for over one to five years. The average size
of loans grew from NT$4,130 in 1961 to NT$15,755 in 1968.

The outstanding amount of loans on December 31, 1968 was
NT$1,243 million. Since all loans are made on the basis of the bor-
rowers' production potential rather than on collateral security and are
closely coordinated with extension services, the record of loan repayment by farmers on maturity has been excellent, being always more
than 95 per cent at any given time.

The capital reserves generated by the program kept in the credit
departments of the participating farmers' associations reached NT$338
million in December, 1968. At this time, the share-capital subscribed
by the farmer-borrowers under the program amounted to NT$107
million. This financial growth has enabled the farmers' associations to
assume a greater share of the responsibility for financing agricultural
development in Taiwan. The improvement in their financial condition has made it possible for the credit departments of the township
farmers' associations to provide more than 60 per cent of the total
farm credit in the province.

The contribution of loan funds for the project operation by the participating farmers' associations out of their own resources steadily
increased as the program gradually expanded. Whereas their contri-

butions were only 20 per cent of the total when the program first started in 1961, the situation was reversed at the end of 1968 when the farmers' associations were growing stronger and fast approaching the goal of sound financial operation and management.

JCRR has maintained a task force of field agents to render supervisory services and technical assistance to the participating farmers' associations in carrying out the program. Besides the on-the-job training provided by its fieldsmen, JCRR has also conducted training lectures and refresher courses in cooperation with the provincial and local agencies concerned. The training courses are designed primarily for new participating farmers' associations to familiarize their chairmen, general managers, credit chiefs, credit men, extension men, and chief accountants with the procedure and methods for implementing the program. The refresher courses are conducted from time to time for old participating farmers' associations to review and evaluate their operation of the program with a view to its further improvement. Cost analysis is one of the most important subjects at the refresher courses, because though the deposits and loans handled by the farmers' associations are small in size, the supervised credit operation is expensive at all times.

Cost reduction, as shown in Table 19-2, has made it possible for the farmers' associations to lower their interest charges on loans to their members, on the one hand, and to acquire a sizeable amount of net earnings from the credit operations every year, on the other.

Table 19-2. Cost of deposits and loans under the unified credit program

Year	Number of FAs investigated	Monthly cost of deposits (%)			Monthly lending overhead (%)	Total monthly cost (%)
		Interest	Overhead	Total		
1962	113	0.79	0.23	1.02	0.19	1.21
1963	167	0.79	0.20	0.99	0.16	1.15
1964	198	0.64	0.16	0.80	0.16	0.96
1965	225	0.63	0.18	0.81	0.16	0.97
1966	237	0.60	0.18	0.78	0.16	0.94
1967	249	0.55	0.18	0.73	0.16	0.89
1968	259	0.53	0.20	0.73	0.16	0.89

Source: Agricultural Credit Division, JCRR.

At village meetings, programs were held to train farmers and by means of visual aids explain the advantages and responsibilities of membership in a permanent cooperative credit system. This measure helped strengthen the credit departments of farmers' associations and cut down the number of demands for reimbursement of the farmer-borrowers' share-capital investment in the credit departments when their loans were repaid.

20. Rural Health and Family Planning

Soon after the transfer of JCRR headquarters to Taiwan in 1949, the Rural Health Division, as one of its first acts, established a health station for each of the 356 townships. Local manpower and resources were mobilized and the local community encouraged to help itself. Assistance was given chiefly in the form of medical supplies, in addition to a small cash grant for travel and per diem expenses to selected health organizations during the initial years of their establishment. When the Taiwan Provincial Government recognized the importance of the health stations to the rural people, it instructed the county and city governments not only to give full financial support to the health units but also to operate them, beginning in 1952.

To standardize the health stations, the township offices were encouraged to construct and furnish new buildings, according to designs drawn up by JCRR. JCRR subsidized one third to one half the cost of construction. Since 1961 the health stations have been further decentralized by being incorporated with township offices.

The next effort was the rehabilitation of the water works. Of the 127 in the rural districts, 32 had been damaged by Allied bombing beyond repair, 77 needed immediate restoration, and only 18 were in fairly good condition and still usable. As a first step JCRR appropriated the equivalent in local currency of US$180,000 for the rehabilitation of the damaged water works. In the second year the Provincial Government earmarked the equivalent of US$450,000 for the purpose; the JCRR contribution was reduced to one third of the construction cost and the benefited communities made up the remaining one third of the expenditures. In the third year of the program operation the Provincial Government increased its water works budget to one million US dollars, and the benefited communities matched it with an equal amount. Since this sum was all that was required to undertake water works repair and rehabilitation, JCRR assistance was discontinued.

Innovative-Type Projects

Rabies control, another health program, was started with a pilot project in 1956 in Yunlin County, which was a highly endemic area. A newly available avianized rabies vaccine was imported from the United States. Since the new product was an attenuated living virus vaccine that has to be kept refrigerated, a cold storage room was constructed and thermos jugs for field use were purchased. The vaccine was ordered from America with specific instructions that the shipment must arrive by Christmas time, so that inoculation could be completed during the cold weather, before the Chinese Lunar New Year in late January of 1957. To carry out the project, village clerks were mobilized to take a census of the dogs in every village, and veterinarians were employed to inoculate the animals. The staff members of the health stations were assigned the tasks of keeping records, preparing the vaccine, and putting tags on dogs immediately after inoculation.

Control teams composed of village clerks, veterinarians, and health personnel went from village to village to do the inoculation. The vaccine was kept in thermos jugs filled with ice. All adult dogs (except pregnant bitches) in the county were inoculated within a period of ten days. Thereafter any stray dog without a tag on it would be caught and killed. Not a single case of rabies was reported in the next twelve months in the areas covered by the project. With Yunlin County as an example, the program was extended to other counties and cities in the south in the second year and to those in the north in the third year. At the completion of the program in the fourth year, rabies had been completely eradicated.

The Family Planning Program

Although the importance of family planning was early recognized by JCRR, nothing could be done about it until the time was ripe. As a first step, the China Family Planning Association was organized in

August, 1954. With the financial and technical assistance of JCRR and annual appropriations from the Provincial Government, the Association launched an island-wide family planning education program, under which the use of contraceptive methods was promoted along with first aid training for women. The amount of annual provincial subsidy to the Association rapidly increased from US$1,250 in 1956 to US$17,500 in 1960 in local currency. JCRR assistance to the Association also increased in proportion up to the latter year when the Provincial Health Department assumed the financing of the Association.

The need for family planning in Taiwan was highlighted by the population increase from 1953 to 1958, during which period the natural rate of increase (i.e., the difference between the birth and death rates) averaged about 3.5 per cent a year, culminating in a grand total of ten million in 1958. This meant that 350,000 new babies were born every year. To feed and raise the newcomers, as Chiang Monlin, then JCRR chairman, had pointed out,[1] there would be required annually: (1) 52,500 metric tons of brown rice costing NT$199.5 million; (2) 7,000 new classrooms in primary schools built at a total cost of NT$420 million and staffed with 10,500 new schoolteachers at a total salary of NT$75.6 million; and (3) a minimum of NT$70 million in clothing. Because of his prestige as JCRR chairman and veteran educator, Chiang's words profoundly influenced public opinion on the burning question of family planning.

As public opinion gradually veered in favor of family planning, the Provincial Government established in 1959 a prepregnancy clinic in each of the public hospitals and health stations with JCRR assistance. As a result, in the course of the next few years, one third (120 in all) of the township health stations employed a full-time female worker to push the program.

In 1962 the Population Council of New York began to take an active interest in the population question in Taiwan. With its financial assistance and the technical assistance of the University of Michigan, the Provincial Health Department established the Taiwan Population Studies Center at Taichung to carry out demographic studies and to evaluate the family planning action program. In the meantime, the

---

[1] Chiang Monlin, *Taiwan's Increasing Population: A Pressing Problem,* pamphlet (Taipei: JCRR, 1959).

Provincial Maternal and Child Health Institute took up the family planning action-cum-research program, which resulted in the adoption of the intra-uterine loop device by the medical and health authorities. In this work the Institute had the benefit of invaluable services rendered by the Population Council and the University of Michigan.

The family planning program, which had been in progress under the sponsorship of the Provincial Health Department, received a shot in the arm when JCRR obtained for it from the United States counterpart fund a pledge of financial assistance amounting to the equivalent in local currency of US$1,500,000 for a five-year period beginning in 1964. The administration of the medical aspects of the program has been assigned to the China Maternal and Child Health Association, which was organized for the purpose.

In 1966 the family planning program was introduced into the armed forces, with the Surgeon General's Headquarters responsible for the work among new military recruits and the Chinese Women's Anti-Aggression League for that among the military dependents.

With the assistance of the Population Council of New York, the University of Michigan, and JCRR, the Provincial Department of Civil Affairs has been publishing an annual demographic fact book, which gives information on the annual birth and death rates, the age-specific fertility rates of women and married women of child-bearing age, and other relevant items by province, counties, cities, and townships. The facts thus made available together with the annual surveys of contraceptive users conducted by the Population Studies Center form the basis for evaluation and for guiding the action program.

The cumulative result of years of effort by the provincial authorities, public and private agencies, and JCRR was the promulgation by the Central Government of a set of Regulations Governing the Implementation of Family Planning in Taiwan on May 22, 1968. These Regulations recognized the need for promoting national health and raising the standard of family living by maintaining the natural annual growth rate of population in Taiwan at a reasonable level. They permitted the people to practice birth control of their own free will by the use of contraceptive methods. They provided that married women may go to a public health or medical agency for examination before pregnancy or childbirth, for guidance on maternal and child health,

on the regulation of conception, and on post-pregnancy health, either
free of charge or by paying a reduced fee for such services. They
further provided that a married woman with three children or more
may, of her own free will, ask a public health or medical agency for
advice on contraceptive methods, which would be provided either
free of charge or at reduced rates for those from needy families.

Aside from these basic principles, the Regulations also provided
administrative agencies, budgets, and staff for the population and
family planning program. This provision represented an immense forward step. Prior to this time, the natural increase rate had been already lowered from 35 per thousand in the 1953-1958 period to 23 per
thousand in 1967. It is expected that, with the official promulgation of
a clear-cut governmental policy in the form of the Regulations just
mentioned, more progress will be made in the years to come.

Rural-Health Education and Sanitation

One village was selected as a model by JCRR to demonstrate rural
health and sanitation measures. All the villagers were required to
participate in a program that had been carefully prepared so that they
could carry it out for themselves, with the means at hand.

Habits of personal hygiene were stressed. Each member of a family
was required to have his own toilet articles—toothbrush, cup, wash
basin, and towel—and was instructed to wash his hands before meals
and after going to the toilet.

Home sanitation was emphasized. Families were required to protect food with screen covers, and water containers with wood covers.
Cupboards and shelves were provided. Latrines were covered. Houses
and furniture were scrubbed, and walls and doors painted. Fowls were
confined in pens or cages. Pigsties were scrubbed. Insecticides were
used against flies and cockroaches. Flowers were planted.

Family planning was taught. The women were taught that too
frequent and too close childbirths would age them prematurely,
whereas child spacing would enable them to retain their physical at-
tractiveness longer and let them and their children enjoy better health.

When the program had succeeded in one village, it was extended to
others and even to the offices of local government organizations and
voluntary agencies, railway and bus stations, theaters, and restaurants

in the township seats. Later on, other undertakings were initiated, such as the paving of roads with asphalt, the laying of drainage ditches, and the construction of a simple piped-water supply system, with the village people sharing one half the cost.

Miaoli County may be taken as an illustration of the program. The county, composed of 18 townships, 263 villages, and 76,620 families with a total population of 491,806 as of September, 1967, was chosen as an experimental area for this purpose in October, 1965. By the end of 1967, the rural sanitation program had been completed in all 18 townships. This program covered sanitation improvement of government organizations and public places in the township seats, and health education and home sanitation improvement in one village for every township. Up to the time of writing in the summer of 1968, each of the 18 township offices in the county had carried out the program in from one to four villages by their own efforts, making a total of 31 villages that had completed the program. Of the 19 villages which had completed some of the engineering items, five had their piped-water works constructed, eleven had 22,404 square meters of their roads paved with asphalt, fifteen had 36,311 meters of their drainage ditches lined with cement, ten had 10,810 square meters of their floors paved with cement, seven had 90 new water wells dug, thirteen had 82 of their latrines remodeled, and six had 691 square meters of new ceilings installed. All this was done at a cost of US$124,353 in local currency, of which the village beneficiaries themselves contributed 56.7 per cent in kind, cash, and labor, the township offices 15.3 per cent, and the Miaoli County Government and JCRR 14 per cent each.

International Technical Cooperation

In carrying out its rural health program in Taiwan, JCRR has been fortunate in having the technical cooperation of the World Health Organization (WHO), the United Nations Children's Emergency Fund (UNICEF), the United States AID Mission to China, and the Rockefeller Foundation. Two of the successful health projects implemented in cooperation with international agencies are malaria eradication and tuberculosis control.

Malaria had been highly endemic in Taiwan prior to its retrocession to China at the end of World War II. As a first step toward controlling the disease, the Taiwan Malaria Research Institute was established by the Ministry of Health in cooperation with the Rockefeller Foundation in 1946. After the fall of the Chinese mainland in 1949, however, the Rockefeller Foundation discontinued its support to the malaria program in Taiwan and JCRR had to assume the responsibility of helping the Taiwan Malaria Research Institute carry on its malaria studies.

The Institute initiated a pilot control project in two townships of southern Taiwan in 1952. It was aimed at killing the vector mosquito, *Anopheles minimus,* by spraying human dwellings and stables with DDT. The program soon aroused the interest of the World Health Organization which, in response to a request by the Chinese Government, dispatched a three-man team of malaria experts to help in the planning. With sizable aid from the United States beginning in 1953, the pilot project was gradually expanded into a province-wide four-year multilateral malaria control program, in which the Provincial Government, the MSA Mission to China, the Council for United States Aid, and JCRR took part.

JCRR assistance was rendered chiefly in the form of subsidies to pay for travel and per diem expenses for military malaria liaison officers who were working at the headquarters of the civilian malaria program so that the military and civilian malaria control programs could proceed simultaneously under unified technical direction. As the other aid-giving agencies had increased their contributions to meet all the budgetary requirements and extended the period of their assistance in order to transform the program from one of control to one of eradication, JCRR financial assistance to it was terminated after 1956. After many years of vigorous effort by all parties concerned, the disease was completely eradicated; WHO officially pronounced Taiwan to be free of malaria in 1965.

The tuberculosis control program in Taiwan began with a JCRR grant to the Provincial Taipei Tuberculosis Center late in 1949 for implementing a pilot project for the establishment of a BCG producing laboratory, tuberculin testing of school children in Taipei City and its vicinity, BCG vaccination of the negative reactors, and X-raying of the positive reactors. In the next year the United Nations Children's

Emergency Fund participated in the project by providing vehicles for the BCG teams, additional equipment and supplies, and fellowships. The three-man team of TB experts, who arrived in May, 1951, helped train and organize 23 local BCG teams. Tuberculin testing and BCG vaccination were soon extended to the whole province. In the meantime JCRR continued to make grants to support local BCG teams and for the remodeling or construction of buildings for the BCG laboratory, tuberculosis sanatoria and centers, and dormitories.

JCRR was an active promoter of the Taiwan Tuberculosis Association, which was organized in 1953 and transformed in 1958 into the National Tuberculosis Association of the Republic of China. The Provincial Government made an initial contribution of US$450,000 worth of local currency to the Association and, when this sum was used up, annual grants to it of US$25,000.

A special grant was made to the Provincial Taipei Tuberculosis Center in 1956 to undertake a mass X-ray examination of chest and pulmonary tuberculosis ambulatory chemotherapy of the aboriginal people. In the next year the pilot project was expanded into a province-wide tuberculosis control program jointly sponsored by the Provincial Government, the UN Children's Emergency Fund, WHO, United States aid agencies, and the Taiwan Tuberculosis Association. Since then, JCRR has discontinued its support for the program.

With the technical and financial assistance of JCRR, a cooperative project for mass sputum examination and chemotherapy of open cases has been jointly implemented by the Provincial Taipei Tuberculosis Center and the Health Center on Quemoy island since FY1963. For indigent areas not easily accessible to X-ray vans, this method has turned out to be a practical and economical one.

As a result of the multilateral control program carried out since the early fifties in Taiwan, the TB mortality rate dropped from 285 deaths per 100,000 people in 1947 to 36 in 1966. A reduction in the prevalence of the disease was also reported by the Taiwan Provincial Tuberculosis Bureau, which made province-wide X-ray and sputum surveys of people aged ten or over selected at random in 1957–1958 and in 1962–1963, and found that the percentage of open cases who carried tubercle bacilli in their sputum dropped from 0.7 per cent to 0.5 per cent during the five-year interval between the surveys. The same was true of both advanced pulmonary tuberculosis and tuberculous cavities which dropped, respectively, from 1.6 to 1.1 per cent and from 0.6 to 0.3 per cent during the same interval. But the incidence of suspected pulmonary tuberculosis remained unchanged at 3.5 per cent.

21. JCRR Assistance to Quemoy

Though activities in the first few years were largely confined to the island of Taiwan, the Joint Commission gradually expanded the scope of its work to include the offshore island of Quemoy.

Quemoy, an island just off the coast of Fukien which in the early fifties was sandy and barren, has acquired new importance because of its strategic position in the current struggle between communism and democracy. Aside from Taiwan and the Pescadores, the little group of islets known as Quemoy and a still lesser group to the north of it by the name of Matsu are the only pieces of free soil which the Chinese Communists have not been able to occupy in spite of repeated attempts to capture them in the last twenty years.

After repulsing the successive military offensives launched by the Chinese Communists in 1949, 1954, 1958, and finally in 1960 (during the visit of President Dwight D. Eisenhower to Taiwan), the local authorities on Quemoy, both military and civilian, have become more convinced than ever of the need to improve the economic situation of the islanders by increased agricultural production. For this purpose JCRR was asked to help. Actually, it had started a few projects on Quemoy as early as 1952, especially for plague control and reforestation. As a result, the bubonic plague had been brought under control, and thirty-four million seedlings had been planted; of these, twenty-five million have grown into tall trees that serve as windbreaks and as shady covers for the network of paved roads connecting all parts of the island.

New crops were introduced, pesticides tested, and experiments made with fertilizers and manure. A program of land reform patterned after that in Taiwan was carried out, and the local farmers' associations were reorganized and strengthened. Four-H clubs were introduced for farm youth, home economics courses were provided for farm girls and women and extension services for adults.

These projects and many others were initiated in the fifties. Some of them have been completed and others that are still needed are continuing. Since the developments up to 1961 have been presented in detail in the writer's *Agricultural Development on Taiwan since World War II,* additional data will be given in this chapter to bring the information up to date.

A decade ago, only a few corn plants could be found on the offshore island. But now corn has become one of the favorite field crops of the local farmers. More than 10,000 *mow* (15 *mow* equal one hectare) have been planted to hybrid corn, and the 1967 yield was over 2,500 metric tons. The rapid increase of corn production has stimulated the development of the swine and poultry industry on Quemoy.

The yields of other main crops have all shown increases. The average annual yield of sweet potatoes in 1965–1967 was 2.4 times that in 1952–1954. The production increase of peanuts, barley, kaoliang, and vegetables over the same period was 1.3, four, ten, and sixty times, respectively.

In view of the high prices of fruits on Quemoy, which had to import them from Taiwan, a project for the introduction, trial planting, and seedling propagation of fruit trees was initiated; it was found that guavas, papayas, grapes, oranges, and mangoes could be grown under local conditions. Altogether some 130,000 such fruit trees have been successfully raised since 1961 and some have begun to bear fruit.

As to animal husbandry, an artificial insemination program has been in progress since 1960 for the production of three-way-cross pigs. An integrated hog production project patterned after that of Taiwan was initiated in 1964, in which year a feed mill was also established by the Quemoy county farmers' association to provide the farmers with the necessary balanced mixed feed. The number of farm families participating in the integrated project rose to 400 in 1967. They were particularly interested in the new method of methane generation, which formed part of the project, because it helped them solve the acute problem of fuel shortage on Quemoy. But when the hog population reached 38,384 head in 1967, the overproduction caused a drop in the price of pork. In order to maintain a proper balance between supply and demand and stabilize the pork price, the county farmers' association has since been assigned the responsibility of marketing all hogs on the island.

To further develop irrigation on Quemoy, an Irrigation Shallow Well Fund was established in cooperation with the Quemoy county government in 1964. By 1967 a total of 3,330 additional wells had been dug with JCRR assistance. Some farmers were enterprising enough to build their own. Furthermore, the Taihu Reservoir, with a storage capacity of 1,140,000 cubic meters, was completed in 1967. Water from it is used for the irrigation of 145 hectares of upland crops. These hectares are now being consolidated into uniform-sized plots and at the same time the irrigation management is being improved. This large-scale project will be the first to make full use of the limited water resources on the offshore island. A second project for the construction of a reservoir at Chunlin with a storage capacity of 270,000 cubic meters was begun in October, 1967. When completed, this new reservoir will irrigate through a pipe line system 30 more hectares of upland crops.

The first waterworks on Quemoy began to supply water for a township of 10,000 inhabitants in July, 1965. It was built at a cost of NT$3,280,000, of which JCRR contributed NT$1 million in the form of a grant. A second water works, constituting part of the Quemoy Rural Water Supply Program supported by JCRR, was completed before the end of 1967, benefiting a rural community of 1,600 people.

The services performed by the farmers' associations on Quemoy include the sale of chemical fertilizers, farm implements, seeds, pesticides, feed, and drugs for hog disease control. The amount of business handled by the associations from 1963 to 1967 is shown in Table 21-1.

The agricultural extension educational program on Quemoy covers all five townships, to each of which is assigned one extension adviser for 4-H clubs, one for home economics, and one for adult farm extension. There are altogether 204 Four-H clubs (with 2,377 members), 47 home economics discussion groups (with 639 members), and

Table 21-1. Sales of farm supplies by the farmers' associations on Quemoy

Year	Fertilizers (NT$)	Feed (NT$)	Farm tools (NT$)	Pesticides (NT$)	Seeds (NT$)
1963	4,245,360	4,663,000	90,000	684,000	165,000
1964	4,609,213	2,949,000	60,000	654,000	492,000
1965	6,154,773	3,057,000	62,000	508,000	149,000
1966	5,945,680	1,915,000	207,000	724,000	100,000
1967	5,652,880	3,524,000	270,000	702,000	190,000

92 adult farm extension discussion groups (with 1,318 members). The Quemoy county farmers' association, which oversees the work of the five township farmers' associations, has two supervisers to assist the township extension advisers in carrying out their extension programs.

County and township farmers' associations on Quemoy have been encouraged to finance their own extension programs by expanding their businesses and rendering better services to the farmers. The financial contributions made by the associations have been steadily increasing, as shown in Table 21-2.

Table 21-2. Financing the extension programs

Year	Local contributions (NT$)	JCRR support (NT$)	Total extension expenses (NT$)
1964	208,000	312,000	520,000
1965	247,000	302,000	549,000
1966	270,000	270,000	540,000
1967	276,000	270,000	546,000

The status of the credit services rendered by the Quemoy county farmers' association is shown in Table 21-3.

Table 21-3. The credit services of the farmers' associations

Year	Deposits (NT$)	Loans (NT$)	Number of loans extended	Profits earned (NT$)
1965	477,000	3,431,000	2,157	367,000
1966	1,907,000	3,480,000	1,820	449,000
1967	1,393,000	3,032,000	1,400	613,000

Of the 9,535 farm families on Quemoy, only 4,133 were enrolled as members in the township farmers' associations in 1953, but today the number has increased to 6,575. In other words, over two thirds of the farm families on the island enjoy the privileges of membership in the farmers' associations.

JCRR has assisted the fishermen of Quemoy in constructing powered fishing craft, purchasing fishing gear and supplies, and building a fish market and a refrigeration plant. There are now 85 powered sampans and ten bigger powered boats. The annual fish production increased from 395 metric tons in 1954 to 3,071 metric tons in 1967. However, it is still slightly below the quantity required for consumption by the civilian and military population of the island.

JCRR has also helped revive oyster culture on the coasts of Quemoy by providing concrete posts to serve as culch. This has resulted in an abundant supply of fresh oysters on the local market.

Altogether 341 projects were carried out on Quemoy with a total grant of NT$95,615,000 in the period from FY1952 to FY1967. Through close cooperation with local people, Quemoy has been transformed from a barren island into an attractive, prosperous, and strongly defended one.

22. International Technical Cooperation

The progressive development of agriculture in Taiwan in the late fifties and early sixties led foreign observers to ask if it would be feasible for Chinese technicians to help other countries develop their agriculture. Through long discussions and exchange of views, the first agricultural technical assistance program to a friendly country, the Republic of Vietnam, was launched in 1959 and was followed two years later by one to Liberia.

Since then, many other countries have invited Chinese agricultural teams and missions to help them improve their agriculture. There were in 1968 twenty-six Chinese agricultural teams working in eighteen countries in Africa and seven countries (Vietnam, the Philippines, Brazil, the Dominican Republic, Chile, Saudi Arabia, and Iran) in other parts of the world. They have introduced Taiwan's crop varieties and cultural practices into these countries, and have demonstrated that such varieties and practices are adaptable to the local conditions in the host countries, which are somewhat similar to those in Taiwan.

Although JCRR has participated actively in the assistance programs for all these countries, only a few will be mentioned here to illustrate the specific approach and methods used by the Chinese specialists. The leaders of teams are all well trained agriculturists with years of field experience; the team members are usually graduates of agricultural vocational schools who have worked on farms themselves and in general are of high quality. Discipline and morale are excellent. The teams live on a farm or in a nearby village. Their relations with local people are cordial.

The projects are usually carried out in three stages: survey and planning, demonstration, and extension.

Chinese Teams in Africa

During their survey trips to eighteen countries in East and West Africa, Chinese specialists have found that many field crops, fruits, and trees there are similar to those in Taiwan, such as rice, sugar cane, tea, cotton, corn, sorghum, tobacco, cassava, bananas, pine-apples, papayas, guava, coffee, coconuts, citrus fruits, mangoes, bamboo, sisal, acacia, kapok, and teak. But these countries are more fortunate than Taiwan in that they are free from typhoons and earthquakes. However, lands in many places lie idle and agriculture has yet to be developed. People living in the jungles lead a very primitive life.

In 1960 seventeen African countries became independent and joined the United Nations. But political independence did not bring with it prosperity. A number of government officials of the Republic of China, including H. K. Yang, a delegate to the United Nations in 1948–1959 and a member of the United Nations Visiting Mission to West Africa Trust Territories, 1952 and 1955, made the acquaintance of some African leaders and discussed with them the possibility of helping the agricultural development of their countries. Economic Minister Yang Chi-tseng, who made his first trip to Africa in January, 1960, with Yang, then director of the African Department, Ministry of Foreign Affairs, supported the idea of extending agricultural technical assistance to African countries.

A three-man team headed by Stephen Tolbert, Secretary of Agriculture and Commerce of Liberia, visited Taiwan in March, 1961, and noted that, in spite of a large population and limited land area, Taiwan was able to produce enough rice not only for domestic consumption but also for export, while Liberia, with a much larger territory, did not even produce enough to feed her people and had to import quantities from abroad. The visitors also noted that Quemoy, the small island lying close to the Chinese mainland and within easy range of Communist gunfire, had had its agriculture well developed in the short space of ten years.[1] Later, at the request of the Liberian government, a group of eight Chinese agriculturists headed by a JCRR specialist was organized and sent to Liberia to study agricultural conditions.

Acting on the recommendations made by this group, a farming demonstration team was dispatched to Liberia by the Government of the Republic of China following the conclusion of a bilateral agreement between the two countries. The team, which arrived at Gbedin in November, 1961, consisted of 16 members led by Chow Mai, a horticulturist. Working with local farmers in the demonstration fields, the team achieved very satisfactory results. The varieties of rice and vegetables used for demonstration were well adapted to Liberia and the yields were high. In the following years, the Liberian farmers who had worked with the Chinese team and other farmers in neighboring villages adopted the Taiwan methods of growing rice and vegetables on their own farms with equally good results.

Similar agreements have been signed between the Government of the Republic of China and eighteen other African countries as shown in Table 22-1.

As the number of Chinese agricultural teams in Africa increased, the Chinese Government established in 1961 the Sino-African Technical Cooperation Committee to handle all related matters. The Committee is composed of representatives of the Ministry of Foreign Affairs, the Ministry of Economic Affairs, the Council for International Economic Cooperation and Development, JCRR, and the Taiwan Provincial Department of Agriculture and Forestry, with Vice Foreign Minister H. K. Yang as Chairman. JCRR has been responsible largely for surveys, program planning, and training of Africans in Taiwan.

The action program mapped out by the committee contains measures (1) to invite African leaders and technicians to visit Taiwan and to send Chinese officials and experts to visit Africa with a view to promoting mutual understanding and to paving the way for agricultural technical cooperation; (2) to organize seminars on farming

[1] Sino-African Technical Cooperation Committee, *Sino-African Technical Cooperation* (Taiwan: Secretariat, Republic of China, 1968).

Table 22-1. Technical missions sent to Africa

Country	Category	No. of mission members, May 31, 1968	Date of establishment
1. Botswana	Agricultural	12	Feb. 1, 1968
2. Cameroon	Agricultural	34	Nov. 7, 1964
3. Chad	Agricultural	41	April 17, 1965
4. Congo-Kinshasa	Agricultural	41	Aug. 12, 1966
5. Dahomey	Agricultural	60	March, 1963; withdrawn April, 1965; resumed Oct. 11, 1966
6. Gabon	Agricultural	50	Oct. 23, 1963
7. Gambia	Agricultural	18	June 11, 1966
8. Ivory Coast	Agricultural	160	March 15, 1963
9. Liberia	Agricultural	30	Nov. 28, 1961
10. Libya	Agricultural	6	March 4, 1962
11. Madagascar	Agricultural	15	Dec. 19, 1966
12. Malawi	Agricultural	32	Dec. 24, 1965
13. Niger	Agricultural	42	July 27, 1964
14. Rwanda	Agricultural	31	Jan. 30, 1964
15. Senegal	Agricultural	26	April 29, 1964
16. Sierra Leone	Agricultural	50	June 15, 1964
17. Togo	Agricultural	33	Aug. 6, 1965
18. Upper Volta	Agricultural	34	April 15, 1965
19. Chad	Veterinary	4	June 23, 1967
20. Ethiopia	Veterinary	6	Aug. 14, 1963
21. Ivory Coast	Seed Multiplication	14	April 3, 1968
22. Rwanda	Sugar Mill	6	May 29, 1968
Total		745	

Source: Sino-African Technical Cooperation published by Secretariat, Sino-African Technical Cooperation Committee, Republic of China, June 1968.

techniques for African agriculturists; and (3) to dispatch to African countries farming demonstration teams.[2]

In line with the first measure, up to May, 1968, the Africans (excluding seminar participants indicated below) from 38 countries who had visited Taiwan numbered 600, including presidents, cabinet ministers,

[2] *Ibid.*

and agricultural leaders. The number of Chinese visitors to African countries was 286 (excluding members of Chinese farming teams). This exchange of visits promoted mutual understanding and helped speed up the conclusion of agreements whereby technical teams were dispatched to African countries.

Under the second measure, nine agricultural seminars had been organized by July, 1968. Scholarships were granted to 407 African agriculturists from 30 countries to enable them to come to Taiwan to attend the seminars. With specialists of agricultural experiment stations serving as instructors, the African trainees received both classroom instruction and field training. Since each seminar lasted six to nine months, the trainees had the opportunity to get thoroughly acquainted with the entire process of growing a crop from planting to harvesting. All the participants expressed their satisfaction with the seminars and encouraged their countrymen to attend similar ones planned for later years. After their return, they have continued to work in the field of agriculture, applying what they have learned here to the agricultural development of their own countries. Some of them often write to their Chinese teachers for further advice.

As a follow-up measure, the Chinese Government sent the writer together with a number of the seminar instructors and other experts to Africa to conduct an Afro-Chinese Seminar for Agricultural Technical Cooperation at Abidjan, Ivory Coast, July 26–30, 1965. This seminar was designed for those African agricultural technicians who had participated in the seminars conducted in Taiwan in previous years. During the Abidjan seminar, the participants reported on the work they had done in their own countries since their return from Taiwan and the problems which had arisen. Their Chinese teachers answered their questions and helped them solve their problems. To simplify linguistic difficulties, the participants were limited to those from French-speaking countries. Altogether 54 agricultural technicians representing ten African countries attended the seminar. The leaders of four Chinese farming demonstration teams working in Africa at that time also took part in the discussions. The members of the seminar later made field trips to Korhogo and Yamoussoukro to observe the farms operated under the supervision of the Chinese farming demonstration teams. There they saw that two crops of rice and fourteen kinds of vegetables had been grown with the same successful results as those in Taiwan.

The Abidjan seminar recommended that more land should be reclaimed for growing rice and dryland crops. It was also recommended that in the areas with good natural grasses, more cattle should be raised for use in plowing and cultivating the land. Cattle manure should be used for making compost which the soils need badly. Since Africa has more land and fewer people than Taiwan, and some areas have timely and adequate rainfall, direct planting of rice instead of transplanting should be tried. The present method of shifting cultivation should be changed to fixed farming. Simultaneously, the land tenure system should be improved to encourage the farmers to make greater efforts to increase farm production.

Under the third measure, agricultural missions or teams have been organized and dispatched to eighteen African countries (Botswana, Cameroon, Chad, Congo-Kinshasa, Dahomey, Gabon, Gambia, Ivory Coast, Liberia, Libya, Madagascar, Malawi, Niger, Rwanda, Senegal, Sierra Leone, Togo, and Upper Volta). In addition, there are two veterinary missions, one to Ethiopia and one to Chad, one sugar mission to Rwanda, and one seed multiplication mission to Ivory Coast. Altogether, 745 Chinese agriculturists belonging to the 22 missions are now (1968) working in Africa.

Many team leaders have been former directors of the sugar cane farms of the Taiwan Sugar Corporation. They all have experience in field experiments, and in the demonstration and extension of new varieties and crops. The leader and deputy leader of a team speak English or French. They are paid US$400 to $500 per person per month. The team members are paid about US$200 per person a month. They have studied a little English; a few have learned some of the African dialects.

Discipline and morale of the teams are excellent. Effective teamwork has contributed largely to the accomplishments of the teams under primitive and very poor health conditions. Many members have suffered from malaria. They are hampered by lack of farm tools.

After arriving at a project site, the team members first prepare the land for cultivation with the help of Africans. In several countries the preparation has involved clearing dense forests and leveling slopes. The Africans are efficient in clearing the land, but less so in simple farm work such as weeding.

In the first year the teams have always started with growing rice and vegetables. In some places they have also grown sugar cane, corn, sweet potatoes, tobacco, soybeans, peanuts, or sorghum. Many Taiwan varieties are well adapted to the African countries. The teams have introduced not only Taiwan varieties but also Taiwan farm tools such as sickles, hoes, plows, power tillers, sprayers, rice threshers, pumps, and cultivators. They have also taught African farmers how to use pesticides and chemical fertilizers.

The improved methods introduced into eighteen African countries from Taiwan are now in the extension stage. Ivory Coast may be taken as an example. The 14-man Chinese agricultural team sent to that country began with rice demonstration in 1963. They worked with local farmers and grew three hectares of rice. In 1964 they helped farmers near the demonstration farm grow rice. In 1965 the area where farmers used Taiwan rice varieties and cultural methods increased to 93 hectares. The rice yield of these farmers was four or five times that of their neighbors. In view of this remarkable achievement, President Felix Houphouet-Boigny was kind enough to say that the Chinese agricultural team working in Ivory Coast has demonstrated how rich in agricultural resources his country is. For this reason, his government has invited 160 Chinese technicians to extend the improved Taiwan cultural methods of rice to other parts of the country and has asked 14 others to help with a seed multiplication program established in April, 1968. These Chinese agriculturists have been helping their host country in the implementation of a national rice production program which is aimed at achieving self-sufficiency in 1969.

In a few other African countries the rice yield per hectare on the demonstration farms has been even higher than that in Taiwan because the African soil is largely virgin and therefore more fertile than Taiwan soils.

Vegetable production is also very important in Africa. Most of the countries there are short of vegetables and have to import large quantities to meet the local demands. The Chinese teams from Taiwan have succeeded in growing vegetables in demonstrations. Among those being extended are radishes, carrots, onions, garlic, Irish potatoes, cabbage, Chinese cabbage (Paitsai), watermelons, cucumbers, melons, squash, tomatoes, sweet peppers, eggplant, string beans, leeks, spinach, mustard, asparagus, and lettuce.

Scientific farming can be observed on the plantations growing African crops for export such as rubber, coffee, cocoa, oil palms, and sisal. These plantations are owned and operated mostly by foreign nationals who make large profits while the farmers remain very poor and live under primitive conditions. Dissatisfied, these farmers are easily taken in by Communist propaganda. Fortified by their experiences gained in Taiwan, the farming demonstration teams have made a good start in helping their host countries increase farm production, hoping eventually to solve their food shortage problem and better the livelihood of their farmers.

Chinese Mission to Vietnam

The agricultural technical assistance program of the Republic of China for Vietnam was initiated by JCRR in 1959 at the request of the government of Vietnam. Contracts were signed between the Vietnamese Ministry of Agriculture and JCRR, representing the Republic of China, for the program which was to be financed by the United States Operations Mission to Vietnam (USOM/V). The idea of this tripartite cooperation originated with William H. Fippin, director of the Food and Agricultural Division of USOM/V, who had been commissioner of JCRR in 1952–1957. The Communist aggression in Vietnam was similar to that on the Chinese mainland in 1948–1949. It was considered that the JCRR approach and experiences would be valuable to the rural reconstruction program in Vietnam.

Under a series of contracts, three Chinese technical teams on farmers' cooperative associations, crop improvement, and irrigation development were dispatched to Saigon in December, 1959, July, 1960, and November, 1960, respectively. The salaries of the team members ranged from US$400 to $600 per person per month. They worked closely with the Vietnamese and USOM/V personnel but initiated no activities of their own. They provided technical assistance to their Vietnamese counterparts to help them carry out agricultural programs.

The Crop Improvement Team introduced into Vietnam Taiwan's

varieties including soybeans, sweet potatoes, peanuts, sugar cane, onions, garlic, and many other vegetables. Their yields in many instances were 50 to 100 per cent (or more) above those of the local varieties. The Farmers' Association Team helped the rural people organize farmers' associations for agricultural extension, rural credit, and marketing services. The Irrigation Development Team assisted the Vietnamese Government in surveying, planning, and designing irrigation constructions as well as in training local technicians. The work of these teams contributed significantly to the increase of crop production in Vietnam.

To extend their operation, the three teams were amalgamated in July, 1964 into the present Chinese Agricultural Technical Mission (CATM), which now has a Crop and Livestock Improvement Division and a Farmers' Service Division. The membership of the Mission has increased from 30 to 85, of whom 60 are field agents working with farmers in different areas, and 25 are specialists assigned to various agricultural services of the Vietnamese Government to help in laboratory or experimental field work.

The agencies established and improvements initiated by the Chinese teams and the Mission from 1959 to 1968 included the following:

1. *Farmers' Cooperative Associations.* They assisted in establishing 48 district-level FAs and six fisheries cooperatives for demonstration purposes; assisted in training managerial, accounting, and extension personnel for various levels of FAs, totaling 549 persons; promoted the implementation of a hog-corn program while helping construct improved pigsties, compost houses, etc., totaling 2,804 units; and assisted such FA services as farm credit, distribution and sale of chemical fertilizers, etc.

2. *Crop and Livestock Improvement.* They introduced plant materials of improved varieties for experimentation at various research institutions; assisted these institutions in selecting and breeding adaptable varieties and strains and adopting improved cultural practices with the improved varieties, which outyielded the local varieties (10 to 60 per cent for rice, 40 to 110 per cent for soybeans, 20 to 400 per cent for sweet potatoes, and 300 to 500 per cent for yard-long beans); helped establish, through introduction and trial planting, profitable new crops such as watermelons, yellow gold melons, garlic, and onions of

Taiwan origin; and promoted the propagation and distribution of superior breeding stock of hogs and extension of improved hog-raising techniques such as artificial insemination.

3. *Irrigation Development.* They assisted in designing 56 irrigation projects for an area of 24,247 hectares; assisted in screening and reviewing 24 irrigation projects for an area of 20,085 hectares; participated in surveying sites for 145 projects covering 179,848 hectares; and helped supervise the construction of seven dam projects covering 3,552 hectares.

Close Cooperation Between Local Farmers and Team Members

In working with the Vietnamese farmers, the Chinese teams have adopted a few simple rules: (1) show the local farmers the results of the new farming techniques to arouse their interest in learning such techniques; (2) let the farmers do what they have learned and correct their mistakes, if any; and (3) ask the farmers to teach the same methods to other farmers in the area so as to expand the extension program.

These simple rules have proved to be effective. Vietnamese farmers and Chinese technicians became close friends and worked shoulder to shoulder in the fields. There was a high degree of mutual confidence. As a result of the improved farming techniques, the farmers' production increased; higher production meant more income and a better livelihood for the farming people. Some concrete cases may be cited to show how the Chinese technicians helped the local farmers and how the latter appreciated the friendly assistance rendered them.

One day, some farmers in the Hue area secretly advised members of the Chinese team stationed there to wind up their work for the day and go back to their downtown headquarters before four o'clock in the afternoon instead of the usual six. The technicians took the advice and found to their surprise on the next day that Viet Cong had infiltrated the suburban area on the afternoon of the previous day and forced the villagers to participate in a series of meetings. Another touching story goes that the farmers in a nearby village voluntarily renovated a road hundreds of meters long in less than a week to make the village accessible to Chinese technicians in order to benefit by their technical assistance.

A Vietnamese farmer named Nguyen Hoan Huy lived in a thatched hut at Huong Phu Village in the Hue area. He was so poor that the family of seven, including five children, slept on piles of straw. The Chinese technicians came across Farmer Huy when they passed by his hut on their way to an experimental plot. The farmer asked the Chinese technicians whether they could help him if he should succeed in bidding for some communal land, as he had no land of his own. The Chinese technicians said yes, so the farmer bid and won the right to plant 0.1 hectare of land. The Chinese team supplied him with vegetable seeds, particularly those of cauliflower and eggplant which were in great demand in Hue, as well as insecticide and advice. Farmer Huy obtained 45,000 Vietnamese piasters from the sale of the vegetables he had harvested. As he had never had so much money in his hands before, he was completely surprised by the windfall.

Nguyen Van No, another Vietnamese farmer, aged 55, lived at Tan Hiep Village near Bien Hoa, a town 20 kilometers from Saigon. Though his tastes were simple and his wants few, he had not been able to satisfy them with his meager income from less than a hectare of lean farm land before he received technical assistance from the Chinese team in that area. Following the advice of the team, Mr. No installed a methane gas tank fed by pipes from the latrine to utilize farm and household wastes to generate gas for fuel and home illumination. The residue also made an excellent fertilizer for the soil. Technical advice from the team also helped him turn a tract of idle land into pasture by planting it with various kinds of legumes. This enabled him to obtain a loan for the purchase of three cows. He is now running a profitable farm. The cows provide him with milk and more manure to yield fuel gas and fertilize his land. The Chinese technicians also helped him improve his native chicken breed by crossing it with a New Hampshire male, and taught him to grow vegetables. All this not only meant a better diet and more income for the farmer: it also pointed to a change for the better in rural Vietnam.

All the team members sent from Taiwan to render technical assistance to friendly countries are agriculturists, graduates of agricultural schools or colleges with years of practical experience, or young farmers. The team leaders are chosen from the staff members of agricultural institutions and the members are recruited through competitive examinations. Before leaving for their assignments abroad, they are given a short orientation course to familiarize them with the tasks they are about to undertake. In carrying out their duties after arriving in their host countries, they are guided by one simple rule: to work in the field side by side with the local farmers and let them learn by doing. This rule has been followed wherever Chinese agricultural teams have been dispatched. In this way the local farmers soon learn and master the cultural techniques and are in a position to teach other farmers how to raise some particular crop. Thus, the number of local farmers who have learned the cultural practices introduced from Taiwan keeps on increasing until one day there will be so many local talents that the services of the Chinese teams will no longer be needed. In other words, the members of the Chinese agricultural teams have been sent out with the ultimate objective of working themselves out of their jobs.

PART IV. EVALUATION

23. Successes, Failures, and Unfinished Tasks

The accomplishments of JCRR have been made possible by both people and circumstances. It has always been manned by a staff of highly competent Chinese specialists, many of whom have received advanced academic training in the United States and have had long experience both on the Chinese mainland and in Taiwan. They work in a favorable environment, free from bureaucratic control and red tape, and free to innovate and experiment with new methods and ideas. The relatively high pay scale has made it easier for the Joint Commission to recruit technicians to fill any vacancies in its technical divisions.

Its organization and programs are flexible, unencumbered by the rigidities which characterize the general run of governmental programs and procedures. Owing to its binational character, JCRR enjoys a wide freedom of action along technical lines and is not subject to the pressures of special-interest groups.

The chairman is elected by the commissioners and has the responsibility of general supervision and administration, as mentioned in "Division of Labor among the Commissioners" in Chapter 1. JCRR, therefore, has the advantage of both the committee-type organization and a single-headed agency.

It has benefited by the hybridization of ideas and the interaction of viewpoints among a staff composed of both Chinese and American agriculturists working in close association in the development and implementation of the program.

Its approach is one of integrated rural development rather than an aggregation of localized community development programs. It is true that JCRR has undertaken a few intensive rural village health projects, but the community development approach is not an important feature of the overall program. JCRR has worked horizontally across a wide spectrum of needs and interests of rural society and vertically up and down the hierarchy of many agencies and organizations, public and private, which have a contribution to make to total rural progress.

These unique features have enabled JCRR to contribute significantly to the agricultural output, productivity, and rural prosperity of the Province of Taiwan and the outlying islands of Quemoy and Matsu. They have also contributed to the reputation it has earned among Chinese government leaders and the people at large as a technically competent agency, a source of unbiased information, and an authority whose judgment and viewpoints are respected for their impartiality and objectivity.

The Reasons for Failures

Though most projects have succeeded, the Joint Commission is well aware that a few of them have failed. In a review of its projects implemented during the eleven fiscal years of 1957 to 1967, it was found that five of its nine technical divisions considered as unsatisfactory thirty-one of their projects involving NT$27,147,967 in grants and NT$54,877,133 in loans out of a total of 2,733 projects involving NT$1,646,793,129 in grants and NT$982,057,961 in loans. This means that 1.1 per cent of all the projects were unsuccessful, and 1.6 per cent of the grants and 5.6 per cent of the loans appropriated for them.

After a careful study, the five technical divisions reported the following causes of these failures:

1. Limiting factors or circumstances unforeseen at the time of project planning.

2. Poor management by the sponsoring agency of the facilities provided under the project.

3. Lack of supporting facilities or equipment by the sponsoring agency.

4. Demand for products smaller than predicted owing to the availability of cheaper and better similar products or substitutes.

5. Difficulty in expanding foreign market for product.

6. Inadequate financing from sponsoring agency for supporting activities.

7. Lack of technical personnel from the sponsoring agency to direct the project work.

8. Equipment or facilities provided under project not suitable for use.

9. Scale of project activity too large without first establishing its feasibility through a pilot scheme.

10. Project too expensive and elaborate for general adoption.

11. Project based on political or other considerations with questionable justifiability.

12. Inexperience of the aborigines in the formation of farmers' associations.

13. Technical incompetence on the part of the farmers to meet the requirements of the project.

The failure of some projects was caused not by any single factor, but by a combination of factors. The following cases may be cited as illustrative examples.

Introduction of the Peking Duck

In 1954 the first attempt was made to introduce Peking ducks into Taiwan; 300 Peking duck eggs were imported from Long Island, New York, for propagation and extension. The 172 ducklings hatched from the imported eggs were intended to serve as a nucleus for an expanded program of meat duck production.

As the Peking ducks could save the farmers the trouble of raising hybrid ducks as meat producers, they were warmly welcomed at first. But it was soon found that they were ill adapted to local conditions in Taiwan. For one thing, they were too costly to raise. The customary way of raising ducks in Taiwan is to let them search for food for themselves either in fields or in ponds and streams. Many hybrid duck raisers adopt a nomadic type of farming. They hatch F_1 hybrid ducklings in Ilan in the northern part of the island in the spring and transport them to the south, where the young birds are herded northward through paddy fields to pick up random grains left there after harvesting. Since the rice crop ripens a few weeks earlier in the south than in central and northern Taiwan, the ducks will always have free meals on their way as they are slowly driven from the south to the north. When they reach their destination, they are about ready for the market. But the Peking ducks are not as good and active scavengers as are native ducks and are better adapted to confined raising. They grow fast but consume a great deal of feed, which means higher cost. For this reason, the farmers were soon disillusioned and reluctant to make further investments in Peking ducks. This was particularly true of the initial years when the price of feed grains in Taiwan was high and their importation was rather limited.

Quite aside from the question of feed, market conditions in Taiwan also proved to be a stumbling block. As the roasted duck, Peking style, served in the restaurants is only one course among many that go to constitute a dinner, a smaller size is preferred. This meant that the fattened Peking ducks could not compete successfully with the native ducks. Therefore, their extension was made extremely difficult.

The introduction of Peking ducks into Taiwan was a failure, not because of any technical mistakes in the breeding or propagation, but because of the consumer's preference for the native ducks. The project was finally abandoned in 1963.

An Unsuccessful Fruit Cannery

To make full use of the low-quality oranges produced in the Hsinchu area in northern Taiwan, the Hsinchu Fruit Marketing Cooperative, which specializes in the marketing of fresh oranges, undertook to establish a fruit cannery for the canning of orange segments. For this purpose, it received JCRR financial assistance consisting of a NT$12 million loan in 1964 and another NT$5 million loan in 1966. The former loan was for investment in physical facilities and the latter for use as working capital. Construction was completed in early 1965 and the cannery has been in operation since then.

In its four and a half years' existence, the cannery suffered heavy losses totaling about NT$20 million by the end of June, 1969, and was on the verge of bankruptcy. The failure may be attributed to several causes.

First, the cannery was constructed entirely with external financing. The original plan called for the raising of a substantial amount of equity capital from the 8,500 farmer members of the Hsinchu Fruit

Marketing Cooperative. But owing to the depressed market price of oranges in recent years, the plan did not materialize.

Second, there was overinvestment in physical facilities relative to business volume. The fixed capital investment amounted to NT$25 million, which was out of all proportion to the actual volume of business transactions. This means there has been underutilization of existing facilities.

Third, the orange segment cans, which are largely for export to Europe, did not sell well on the export market in the last four years. At the same time, the cannery also incurred heavy financial losses in another line of business, strawberry jam, which had to compete with similar products dumped on the Japanese market by the Chinese Communists. These were unfavorable environmental factors beyond the control of the Hsinchu Fruit Marketing Cooperative.

Fourth, the location of the cannery was ill chosen, with the result that all the raw materials it needed, except oranges, had to be shipped over long distances. To limit its operations to the canning of orange segments alone would not bring in enough business. Moreover, to expand its operations to include other fruits and vegetables would entail difficulties of transportation.

Fifth, the top management lacked both technical knowledge and practical experience in cannery operation. This was true as regards raw materials procurement, product development, plant operation and management, and domestic and international marketing of the products.

Sixth, the organizational setup was so inefficient that if anything went wrong, it was usually impossible to find out who should be held responsible.

Seventh, the merits of a cooperative organization were not fully incorporated into this venture. Cooperative arrangements were used only in the procurement of raw materials and, as most of the raw materials had to be procured outside the Cooperative, whatever advantages the Cooperative offered affected only a small part of the whole business process. Even if fully exploited, it is questionable that the advantages of a cooperative would be worth much in this situation.

In the final analysis, the problem was one of failure in project evaluation or, more specifically, in benefit-cost studies. The returns were overestimated and the costs underestimated. Elements of uncertainty, both internal and external, were not given due consideration. Furthermore the human factor, which is a prerequisite to the successful operation of any business undertaking, was overlooked.

Lessons of the Tapu Reservoir

The Tapu Reservoir in Miaoli County was constructed in 1956–1960. This concrete gravity-type reservoir is 21.4 meters high and has an effective storage capacity of 9,600,000 cubic meters and a life expectancy of fifty years. It provides irrigation water for 1,343 hectares of farm land in the Tapu area by means of a 10.9-kilometer long main canal, 117 kilometers of laterals and sublaterals and two pumping stations. The lands thus irrigated produce two crops of paddy rice and one winter crop a year.

When construction of the reservoir, the main canal, and seven laterals began in 1956, the estimated cost was NT$40 million. But owing to price fluctuations resulting from currency inflation in the next few years, the total engineering cost actually amounted to NT$54 million.

The original plan left the construction of sublaterals to the farmers themselves. However, they failed to do the work, which therefore had to be done by the local irrigation association with additional cost.

The purchase of right-of-way for canals and compensation for the inundated lands in the reservoir area presented another serious problem in project implementation. At first it was thought that the question could be solved by the local irrigation association with its own funds. After protracted negotiations lasting for almost two years, the association had to borrow NT$8 million from the Land Bank to pay for right-of-way and as compensation. In this way, the total project cost went up to NT$62 million.

Aside from the increased project cost, the farmers had to make a sizeable reclamation investment to convert the original dry lands of irregular topography into terraced paddy lands at an average cost of NT$13,500 per hectare. To finance this operation, many farmers drew upon their own savings, but some had to depend upon short-term loans from the Land Bank at an interest rate of 14 per cent per annum.

Still others who obtained funds from the rural credit cooperatives of the farmers' associations paid as much as 16 per cent per annum.

Since this reclamation investment exhausted the farmers' financial resources, the farmers began to default on payments of the regular water fee and the special assessments. As the regular water fee is for the operation and maintenance of the irrigation systems and the special assessments are for the payment of project loans, the defaults had the net effect of cutting off the irrigation association's life blood and making it impossible to repay the project loans on schedule. As a consequence, the NT$20 million loan made by JCRR to the Tapu project from 1956 to 1960 had to be extended from twelve to fifteen years.

Management of farm soils was another point neglected by the Tapu project. Part of the lands in the Tapu area consists of coastal sand dunes, and the nature of the soil posed difficult irrigation problems which were still unsolved in the eighth year of project implementation.

JCRR made a thorough review of the economic aspects of the Tapu project in 1963. A recalculated benefit-cost ratio (from the original 1.96:1 to 1.15:1) showed that the project was still justified. However, the original cost estimate should have included such important items as sublateral construction, payment for right-of-way, compensation for inundated lands, and reclamation cost.

Engineering design and construction are the major part of an irrigation project. The Tapu case showed that the successful implementation of any irrigation project depends not only on engineering design and construction, but also on a sound policy of funding, resettlement and reclamation, soil and crop pattern, irrigation management, and, most important of all, the financial capacity of the beneficiaries.

JCRR irrigation projects undertaken later than the Tapu project have taken these factors into consideration. Direct and indirect expenses necessary for project implementation are carefully calculated to obtain a realistic cost estimate. Project benefits are also studied in detail and compared with the cost figures to arrive at an overall appraisal of economic feasibility. Thus, the engineering and economic soundness of an irrigation project is carefully worked out before it is implemented.

In the light of the Tapu experience, JCRR has also begun to extend reclamation loans to farmers to level their lands for irrigation. Specialists go to the fields to help in soil management and improve the cultivation of irrigated crops. These activities are now considered as an integral part of all irrigation projects. This is a direct result of the failure of the Tapu project, whose painful lessons have been taken to heart.

In the views of the writer, while JCRR cannot totally exonerate itself of its share of responsibility, it is apparent that the failures may be attributed, by and large, to the inherent weakness of the projects themselves and to lack of necessary supporting facilities from the sponsoring agencies. There may be other failures in the future because JCRR undertakes innovative projects that involve risks.

Unfinished Tasks

In previous chapters, especially those in Part III, mention is made of some projects and tasks that have not been finished or remain to be tackled in the future. For example, with the introduction and development of new crops and new crop varieties for further raising agricultural productivity, there have appeared new diseases and insect pests posing a challenge to entomologists and plant pathologists. In fact, in the future development of agriculture in Taiwan, there are both old and new problems to be solved and difficulties to be overcome; it is the responsibility of JCRR to work in close cooperation with government agencies and farmers' organizations concerned to remove all obstacles that stand in the way of further agricultural progress.

The basic problem in Taiwan is the lack of arable land. In the face of ever-expanding population, this physical limitation has resulted in diminishing farm size and continued farm fragmentation. JCRR has been trying to solve this problem by promoting intensive methods of cultivation and joint farming operations to bring the maximum returns from land. Land consolidation and development of marginal slope lands have also been going on in an attempt to stabilize the size of farms and to enlarge the cultivated area.

But under the small farm system, land will remain a scarce factor in agricultural production and it will be difficult to change the farm structure and maintain the scale of farming at an economically desirable level. Furthermore, with the rapid growth of industry in recent years, more and more rural people have left their farms to seek employment in cities, with a consequent rise in the wages of labor. Simultaneously, there has been a steady encroachment on agricultural lands by industrial and housing constructions. This has aggravated the land problem. The best agricultural land in Taiwan must be used for agriculture. Efforts for increasing agricultural production will continue to emphasize the fuller utilization of the available land resources so that land productivity can be raised. Measures for raising the efficiency and productivity of labor are also an urgent necessity.

The high cost of production requisites, especially fertilizers, has kept the income from farming much lower than that from other occupations. This income disparity has had an adverse effect on agricultural production and the economic well-being of the farming population in general. As mentioned previously, the price of fertilizers in Taiwan is much higher than in other countries. But under the present system of international trade Taiwan farm products have to compete with those of other countries and are often forced to sell at prices well below cost. In order to provide the farmers with incentives to increase farm production, the prices of farm supplies must be lowered sufficiently so that there will be an attractive margin left for the producers. Agricultural industries also need fair margins. Systems for subsidizing irrigation, land reclamation and similar farm developments, price guarantee, and the protection of domestic products in connection with imports need also to be considered as incentives to farmers.

In the Republic of China, as in most other countries, agricultural research and extension education are financed by government funds. But owing to financial stringency of the Chinese Government, such work has relied largely upon the support of grants from JCRR in the past. But as funds are limited and the percentage of these grants against loans is gradually becoming smaller and smaller, the question of financing agricultural research and related activities in Taiwan will become increasingly serious as the years go by. However, in order to maintain the growth of agriculture at a desired rate, research efforts not only cannot be slackened but must be further strengthened. The Provincial Government will, therefore, have to assume an ever greater responsibility for such tasks in the future.

While increasing agricultural production continues to be a major goal of JCRR, in the years ahead, more attention must be paid to marketing. In order to increase the exportability of such economic crops as fresh and processed fruits and vegetables, the farmers in Taiwan need to further improve their production techniques and marketing practices. Though great efforts have been made in these directions, the fact that all such crops are grown by individual small family farms makes quality control and post-harvest handling very difficult. This puts Taiwan products at a disadvantage when they compete with similar products from other countries on the international market, a classic example being banana export. Taiwan bananas had dominated the Japanese market for many years until large quantities of the fruit from Central American countries began to appear in Japan for the first time in 1968. The bananas from Central America are better than Taiwan bananas because of their uniform quality, low spoilage, and better packing. Their export to Japan is handled by two large fruit companies which operate big banana plantations and have their own export plans covering production, harvesting, inspection, packing, and transportation.

Everything is well organized and controlled by the companies, in contrast to the poorly organized condition of the small farms in Taiwan. It is feared that if necessary improvements are not made in time, not only bananas, but also many other farm products, will sooner or later lose their established markets abroad.

The importance of research in agriculture cannot be overemphasized. As agriculture in Taiwan has already attained a high level of development, its further growth will be difficult without the support of advanced technology. An intensified and dynamic research program is therefore essential to the continued development of agriculture on this island. In this regard, JCRR will continue to assist the Provincial Government in training agricultural technical personnel, modernizing research facilities, and establishing regional agricultural research institutions in cooperation with other countries. International cooperation in agricultural research will be especially emphasized, as it will enable this country to keep itself abreast of the latest developments in agricultural technology and help it raise the level of its own research. The Asian Vegetable Development and Research Center and a Swine

Research Center are two of such institutions planned to be established in the very near future. If and when these and similar research centers are eventually established, they will not only help to increase agricultural productivity in Taiwan but also be beneficial to other developing countries in Asia, Africa, and a part of Latin America, to which Chinese agricultural teams or missions have been sent.

The Future of JCRR

In spite of the phasing out of United States aid to China, both the Chinese and American governments have felt it necessary for JCRR to continue its useful activities for the further development of agriculture in Taiwan. The two governments agree to maintain its joint character; the United States will continue to assign an American citizen as one of the three commissioners. There are several reasons for this. JCRR has a substantial record of achievements in agricultural development. JCRR is continuing to participate actively in the development of Taiwan, and in addition, it supports many projects of agricultural assistance to other countries, often in association with similar United States projects. Therefore, it is most important for JCRR to continue its very close cooperation with the United States in the field of agriculture, so that Taiwan may share in the technological advances for which the United States is noted and may make available to other countries the results of agricultural research and improvements undertaken in Taiwan.

The exchange of notes between the Chinese and American governments on April 9, 1965,[1] was essentially an extension of the original bilateral agreement signed in Nanking on July 3, 1948. In the notes the two countries agreed to create a Sino-American Fund for Economic and Social Development (SAFED) out of the residual accumulated counterpart funds generated by the sale of United States aid commodities on the local market. By the creation of such a fund, part of which is appropriated for use by JCRR on an annual basis, the United States showed its continuing interest in Taiwan's economic and social development. Under this new arrangement, JCRR, which has been responsible for planning and programming the agricultural

[1] JCRR, *General Report XVI* (Taipei: JCRR, 1965), pp. 127–128; *General Report XVII* (1966), pp. 1–2.

sector of the economic development program in the last eighteen years, will continue as a bilateral organization. But the responsibility for approving the program is vested in the Chinese Government.

According to the terms of the April 9, 1965, exchange of notes, JCRR is to plan its programs and projects for the further agricultural development of the Republic of China with emphasis on research for developing new products and working out new and better methods for raising productivity as well as stimulating the efficient production of new products.

In line with the changes made in its program objectives and operational guidelines in 1963, mentioned in Chapters 2 and 4, and also in view of the present stage of agricultural development in Taiwan, the 1965 SAFED agreement further requires that JCRR plan its future programs to enable the transfer of the usual types of assistance to other institutions and organizations and to encourage them to devote more of their efforts to research aimed at the introduction of new products and processes so as to increase productivity and trade.

For the realization of these purposes, JCRR has been (1) encouraging or carrying out research on the production and marketing of farm crops, livestock, forestry, and fisheries, and encouraging or carrying out research relating to technological improvements in land improvement and reclamation, irrigation installations, and flood control works; (2) encouraging by means of financial and other forms of assistance the production of new products and the adoption of new farming techniques on the basis of the results of current research; (3) helping to remove institutional bottlenecks that adversely affect both the production and marketing of quality farm products, especially those for export; (4) assisting to push research programs on the potentials of foreign markets for agricultural products produced in Taiwan; (5) cooperating with Chinese government agencies in setting up market grades and standards for agricultural exports; (6) assisting exporters in the promotion of foreign markets; (7) rendering assistance to Chinese and foreign investors in agriculture; (8) taking an active interest in measures for keeping the rate of population growth at a reasonable level; (9) supporting such other programs as rural health, agricultural extension, farm credit, farmers' organizations, and research and training as will contribute to general agricultural development and rural community betterment.

JCRR as a Model for Other Aid-Recipient Countries

The Foreign Assistance Act passed by the United States Congress in 1966 made certain suggestions as to how AID could administer its assistance to the rural sector of aid-recipient countries. It encouraged AID to establish, in cooperation with host countries, Joint Commissions on Rural Development. It said in Section 471 of Chapter 7:

(a) The President is authorized to conclude agreements with less developed countries providing for the establishment of Joint Commissions on Rural Development each of which shall be composed of one or more citizens of the United States appointed by the President and one or more citizens of the country in which the Commission is established. A majority of the members of each such Commission shall be citizens of the country in which it is established. Each such agreement shall provide for the selection of the members who are citizens of the country in which the Commission is established who wherever feasible shall be selected in such manner and for such terms of office as will insure to the maximum extent possible their tenure and continuity in office.

(b) A commission established pursuant to an agreement authorized by this section shall be authorized to formulate and carry out programs for development of rural areas in the country in which it is established, which may include such research, training and other activities as may be necessary or appropriate for such development.

(c) Not to exceed 10 per centum of the funds made available pursuant to section 212 shall be available to the President in negotiating and carrying out agreements entered into under this section, including the financing of appropriate activities of Commissions established pursuant to such agreements.[2]

Such a recommendation is evidence enough that Congress is interested in applying the successful experience gained by AID in the

[2] U.S. Senate Committee on Foreign Relations and U.S. House of Representatives Committee on Foreign Affairs, *Legislation on Foreign Relations With Explanatory Notes* (Washington, D.C.: U.S. Government Printing Office, Jan. 1967), pp. 25–26.

Republic of China through the workings of JCRR to other aid-recipient countries.

So far, efforts made by AID directors to set up joint commissions on rural development in the Republic of Vietnam, the Philippines, and a few other countries have not yet borne fruit. But in China, JCRR was organized with the greatest dispatch, with the Government of the Republic of China taking the initiative instead of the United States. Similarly, the initiative for the continuation of JCRR after the termination of United States aid also came from the Chinese side.

The achievements recorded in this book were undreamed of when the Joint Commission was first set up in Nanking in 1948. They are a measure of the soundness of the bilateral approach to the problem of rural reconstruction. Whether the same approach will work in other developing countries with similar results is a question which cannot be answered until it has been actually put to the test in those countries.

APPENDIXES AND INDEX

Appendix I. Exchange of letters between President Harry S Truman and President Chiang Kai-shek relating to the formation of a China-United States agricultural mission, 1946*

THE WHITE HOUSE
WASHINGTON

June 17, 1946

My dear President Chiang:

I am happy to inform you that in response to the request of the National Government of the Republic of China this Government is sending to China a group of eight agricultural specialists under the leadership of Dean C. B. Hutchison to work jointly with agricultural leaders appointed by the Government of China on problems relating to the development of China's agriculture. This mission will be ready to leave for China the latter part of June. A list of its members is attached.

I am pleased that this arrangement is to be carried out because it is my firm belief that any plan for cooperation in economic development between our two countries should include agriculture, the major source of income for such a great proportion of China's population. In the experience of the United States, agricultural improvement has been found so important in promoting security, producing industrial raw materials, providing markets for industrial products, and raising the level of living that we believe a successful national development cannot be assured unless the development of agriculture proceeds

* *Source:* Office of Foreign Agricultural Relations, USDA, *Report of the China–United States Agricultural Mission* (Washington, D.C.: Office, May 1947), pp. i, ii.

simultaneously with the development of other elements in the national economy.

While we hope that our agriculturists on this mission may be able to render substantial service toward the betterment of China's farming, we also are aware that our own agriculture is already indebted to your country for valuable agricultural material which has been introduced into the United States. Moreover, we still have much to learn from Chinese agriculture.

A higher level of living for the whole of China's population, which can hardly be achieved without a strong development of agriculture, is the necessary foundation for the achievement of results that will benefit both of our countries, including an expansion of complementary trade and the development of China's industrial program.

It is in this spirit of sharing in an endeavor of great potential value to our two countries that the American members of this mission are visiting your country. I shall receive with interest the report of this group.

I am asking General of the Army George C. Marshall to deliver this letter in person. I wish to convey with it an expression of my warm personal regards to you and the continuing interest of our people in the welfare of your country.

Very sincerely yours,

[Signed]HARRY S TRUMAN

His Excellency
Generalissimo Chiang Kai-shek,
President of the National Government of the Republic of China,
Nanking, China

[TRANSLATION]

THE NATIONAL GOVERNMENT OF THE
REPUBLIC OF CHINA

July 31, 1946
Nanking, China

My dear President Truman,

It gives me great pleasure to acknowledge receipt of your letter of June 17, 1946, delivered in person by General of the Army of the United States, George C. Marshall. We are very appreciative of the splendid response you have made to the request of this Government for the despatch of an agricultural technical mission to this country.

We have been for centuries primarily an agricultural nation. The farmer is traditionally regarded with affection and respect. During recent times, unfortunately, our agricultural technique has fallen behind due to delay in the adoption and application of new scientific methods. I am keenly conscious of the fact that unless and until Chinese agriculture is modernized, Chinese industry cannot develop; as long as industry remains undeveloped, the general economy of the country cannot greatly improve. For this reason, I heartily agree with you that any plan for co-operation in economic development between our two countries should include agriculture.

I feel highly complimented by your statement that China has contributed some valuable material to your agriculture. I sincerely hope that, through your co-operation, we shall be able to make further significant contributions to this field, for the benefit of mankind.

I congratulate you upon the happy selection that you have made of the personnel constituting your mission. On our part, we have chosen a corresponding number of men of high quality and long experience to work in conjunction with your mission. Already the spirit of co-operation between the two groups is evident. It is my firm belief that the two groups working together will succeed in evolving plans and projects which will prove beneficial to China as well as helpful to the development of economic and trade relations between our two countries.

We are now actively taking up the work of national reconstruction. Agricultural improvement being the foundation of such reconstruction, I assure you that the work of your mission will receive my continued attention and support.

I avail myself of this opportunity to convey to you my warmest regards and highest esteem.

Very sincerely yours,

[Signed] CHIANG KAI-SHEK

His Excellency, Mr. Harry S Truman
President of the United States of America,
Washington, D.C., U.S.A.

Appendix II. Text of notes exchanged August 5, 1948, by the United States and China providing for the establishment of a Sino-American Joint Commission on Rural Reconstruction, 1948 (renewed June 27, 1949)*

Note from Ambassador Stuart to the Foreign Affairs Minister

Excellency:

I have the honor to refer to Section 407 of the China Aid Act of 1948 enacted by the Government of the United States of America (hereinafter referred to as the Act), which provides, among other things, for the conclusion of an agreement between China and the United States of America establishing a Joint Commission on Rural Reconstruction in China. In pursuance of the general principles laid down in the Act, and in particular Section 407 thereof, I have the honor to bring forward the following proposals regarding the organization of the Joint Commission and related matters:

(1) There shall be established a Joint Commission on Rural Reconstruction in China (hereinafter referred to as the Commission), to be composed of two citizens of the United States of America appointed by the President of the United States of America and three citizens of the Republic of China to be appointed by the President of China. The Commission shall elect one of the Chinese members as Chairman.

(2) The functions and authority of the Commission shall, subject to the provisions of the above-mentioned Section of the Act, be as follows:

(a) To formulate and carry out through appropriate Chinese Government agencies and international or private agencies

* *Source:* Mimeographed copy kept in JCRR files.

in China a coordinated program for reconstruction in rural areas of China (hereinafter referred to as the Program);

(b) To conclude arrangements with the agencies referred to in the preceding paragraph establishing a basis for their cooperation;

(c) To recommend to the Governments of the United States of America and of China, within the limits prescribed by the Act, the allocation of funds and other assistance to the Program, and to recommend to the Government of China the allocation of such other funds and assistance as are deemed essential to the success of the Program;

(d) To establish standards of performance for implementation of the Program, including the qualifications, type and number of personnel to be used by cooperating agencies in the Program, and to maintain a constant supervision of all phases of the Program, with authority to recommend changes in or stoppage of any phase of the Program;

(e) To appoint such executive officers and administrative staff as the Commission deems necessary to carry out the Program, it being understood that the chief executive officer shall be a citizen of China. Salaries, expenses of travel and other expenses incident to the administrative functions of the Commission itself shall be paid from funds made available under Section 407 (b) of the Act.

(3) In its Program the Commission may include the following types of activity to be carried out in agreement with the agencies referred to in paragraph (2) (a):

(a) A coordinated extension-type program in agriculture, home demonstration, health and education, for initiation in a selected group of *hsien* in several provinces to include a limited number of subsidiary projects suited to conditions in the areas where the program is developed, in such fields as agricultural production, marketing, credit, irrigation, home and community industries, nutrition, sanitation, and education of a nature which will facilitate the promotion of all projects being undertaken;

(b) Consultation with the Chinese Government concerning ways and means of progressively carrying out land reform measures;

(c) Subsidiary projects in research, training and manufacturing, to be carried out in suitable locations to provide information, personnel and materials required by the Program;

(d) Projects to put into effect over a wider area than provided for in the coordinated extension-type program specified in (a), any of the above lines of activity which can be developed soundly on a larger scale, of which examples might be the multiplication and distribution of improved seeds, the control of rinderpest of cattle, the construction of irrigation and drainage facilities, and the introduction of health and sanitation measures;

(e) Related measures, in line with the general objectives of this Program;

(f) The distribution of the assistance in this Program, on the principle of giving due attention to strengthening rural improvement in areas where selected projects can be progressively developed and where their development will contribute most effectively to the achievement of purposes for which this Program is undertaken, but that the principle of distributing aid will not be controlled by proportionate or geographical consideration *per se.*

(4) In respect of any decision of the Commission, the approval of the Government of China shall be obtained prior to its execution if the Commission or its Chairman, with the concurrence of the Chinese members, deems it necessary.

(5) The Commission shall publish in China and transmit to the Government of the United States of America and the Government of China, in such form and at such times as may be requested by either of the two Governments, full statements of

operations, including a statement on the use of funds, supplies and services received, and will transmit to the two Governments any other matter pertinent to operations as requested by either of the two Governments. The Government of China will keep the people of China fully informed of the intended purpose and scope of the Program and of the progress achieved by the Commission in implementing the Program, including the nature and extent of the assistance furnished by the Government of the United States of America.

(6) The Government of China will upon appropriate notification of the Ambassador of the United States of America in China consider the United States members and personnel of the Commission as part of the Embassy of the United States of America in China for the purpose of enjoying the privileges and immunities accorded to that Embassy and its personnel of comparable rank. It is understood that the Ambassador of the United States of America in China in making the notification will bear in mind the desirability of restricting, so far as practicable, the number of officials for whom full diplomatic privileges and immunities would be requested. It is also understood that the detailed application of this paragraph would, when necessary, be a subject of inter-governmental discussion.

(7) All supplies imported into China for use in the Program shall be free of Customs duties, Conservancy dues and other charges imposed by the Government of China on similar supplies which are imported through regular commercial channels.

(8) The Government of the United States of America and the Government of China will consult with respect to problems incident to the interpretation, implementation and possible amendment of the terms of the agreement embodied in this exchange of notes whenever either of the two Governments considers such action appropriate.

(9) The Government of the United States of America reserves the right at any time to terminate or suspend its assistance, or any part thereof, provided under this exchange of notes. Assistance furnished by the Government of the United States of America

under Section 407 of the Act and pursuant to this exchange of notes shall not be construed as an express or implied assumption by the Government of the United States of America of any responsibility for making any further contributions to carry out the purposes of Section 407 of the Act or of this exchange of notes.

(10) This note and Your Excellency's reply accepting the above proposals on behalf of the Government of China will constitute an agreement between the two Governments in the sense of Section 407 of the Act. Subject to the provisions of paragraphs (8) and (9), this exchange of notes will remain in force until June 30, 1949, or, upon the request of either Government transmitted to the other Government at least two months before June 30, 1949, until the date of termination of the Economic Aid Agreement between the two Governments concluded on July 3, 1948.

I avail myself of this opportunity to renew to Your Excellency the assurances of my highest consideration.

Note from Dr. Wang Shih-chieh to the Ambassador

Excellency:

I have the honor to acknowledge receipt of your note of today's date which reads as follows:

[Full text of Ambassador's note]

On behalf of the Government of China, I have the honor to accept the proposals contained in the note quoted above.

In recognition of the importance of the Program as one of the essential means of achieving the objectives in which the Governments of China and of the United States of America unite in seeking under the Economic Aid Agreement between the two Governments concluded on July 3, 1948, the Government of China undertakes to afford to the execution of the Program the full weight of its support and to direct cooperating agencies of the Government of China, including the local officials concerned, to give such assistance and facilities as are essential to the success of their undertakings under the Program.

I avail myself of this opportunity to renew to Your Excellency the assurances of my highest consideration.

Index

Wilcox, R. F., 191–192
Willson, Clifford H., 20
Windbreaks, 54, 104, 106, 149, 169, 222
Winter crops, *see* Crops, winter
World Bank, 17, 83
World Census of Agriculture, 201
World Health Organization, 218–220
World War II, 52, 54, 59, 91, 109, 113,
    130, 138, 203
Wu Basin, 119–120
Wu Chi, 107
Wuchang, 12
Wuchi-Chosui basin, 84

Wukung, 12
Wushantou Reservoir, 126
Wusheh, 107

Yang, H. K., 228–229
Yang Chi-tseng, 228
Yen, C. K., 26
Yen, Y. C. James, 13, 14, 15, 20
Yenshui Chi Diking Project, 114
Yunlin, 122, 124–125, 214

Zehngraff, Paul, 169